CALIXARENES

Monographs in Supramolecular Chemistry
Series Editor: J. Fraser Stoddart, *University of Sheffield, U.K.*

This series has been designed to reveal the challenges, rewards, fascination and excitement of this new branch of molecular science to a wide audience and to popularize it among the scientific community at large.

No. 1 Calixarenes
By C. David Gutsche, Washington University, St Louis, U.S.A.

Forthcoming Titles

Cyclodextrins
By Fraser Stoddart and Ryszard Zarzycki, University of Sheffield, U.K.

Crown Ethers and Cryptands
By G. W. Gokel, University of Miami, U.S.A.

Cyclophanes
By F. Diederich, University of California, U.S.A.

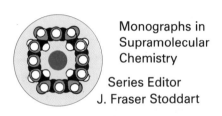

Monographs in
Supramolecular
Chemistry

Series Editor
J. Fraser Stoddart

Calixarenes

C. David Gutsche
Washington University
St Louis, U.S.A.

ROYAL
SOCIETY OF
CHEMISTRY

British Library Cataloguing in Publication Data
Gutsche, C. David
   Calixarenes.
   1. Macrocyclic compounds
   I. Title II. Series
   547'5

   ISBN 0-85186-916-5

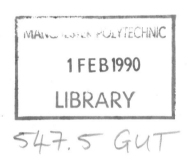
Published by The Royal Society of Chemistry,
Thomas Graham House, Science Park, Cambridge CB4 4WF

Set by Unicus Graphics Ltd, Horsham, West Sussex
and printed in Great Britain

Preface

Science comprises a marvellous mosaic of individual pieces — some large, some small, some of cosmic importance, some of minor consequence. Each has a shape, color, structure, and set of characteristics that distinguish it from all others and make it unique. The present book deals with one such piece which fits somewhere in the middle of these boundaries, being medium in size and modest in importance. But, it is a piece that boasts a more than humdrum history, and it is acquiring an interesting patina with the passage of time.

One important purpose of this book is to provide a complete and timely survey of the chemistry of the molecular baskets called 'calixarenes', a survey that might serve as a springboard for researchers interested in expanding this domain of supramolecular chemistry. It is also hoped that the historical vignettes, the pictures, and the short biographies that are included will underscore the fact that research is a very human endeavor whose course is seldom straight or always upward. In a Memorial Lecture delivered in 1932 Thomas Huxley said that 'we are not in the habit of inspecting the houses in which we live. Except for the historians of science, nobody studies at first hand those contributions to knowledge to which the great discoverers of the past owe their scientific reputation'. The present book attempts to address this problem by taking the reader inside the house of one small chemical family so that future generations who inhabit the house will have at least a modicum of awareness of their scientific ancestry.

The area called 'calixarene chemistry' started modestly in the 1970's with a small handful of players. Slowly gaining momentum in the late 1970's, its pace accelerated in the 1980's and now engages the attention of numerous researchers. A picture and a short biography are included in the book for a few of these individuals, the ground rules for the choices being that the individual is deceased, or has won a Nobel Prize, or has published several papers in the field of calixarene chemistry. These selections are in no way intended to minimize the contributions of the many who have not been singled out in this fashion. Indeed, it is more than likely that some of those not so chosen will become the leaders of the field in the years ahead, and it is anticipated that their accomplishments will be duly noted in future volumes dealing with this topic.

It was my good fortune to be among that small group of individuals who dabbled with calixarene chemistry in the 1970's, for it has been exciting to see the field be planted, cultivated, and harvested. A most significant part of whatever success I have had in helping with this development must be attributed to the very capable group of coworkers who assembled around me during the past fifteen years and to the splendid cooperation of the Petrolite Corporation and its fine chemists, especially Drs. Jack Ludwig and John Munch. Another significant part must be attributed to a particularly felicitious choice of a wife. Alice Gutsche, who has been a part of my research group for most of this period, played a major role in the writing of this book. Her tireless work in compiling a reference library on calixarene chemistry, her expertise as an editorial critic, and her willingness to endure the tedium of proof reading qualify her for coauthorship which, however, she demurs from accepting. To my splendid group of young men and women, whose names will be found throughout the book in references and footnotes, and to my wife Alice I give my heartfelt thanks for their unremitting efforts to put calixarene chemistry on the map. Particular credit is extended to Iftikhar Alam, Janet Rogers, Keat-Aun See, and Donald Stewart for their careful readings of the manuscript of this book. Finally, great appreciation is expressed to the National Institutes of Health, the National Science Foundation, and the Petroleum Research Fund administered by the American Chemical Society for the many years of generous support they have given our research program.

C. David Gutsche
St Louis, Missouri
October 1988

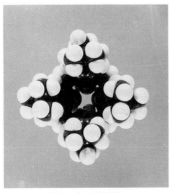

p-*tert*-Butylcalix[4]arene

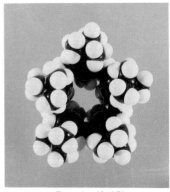

p-*tert*-Butylcalix[5]arene

p-*tert*-Butylcalix[6]arene

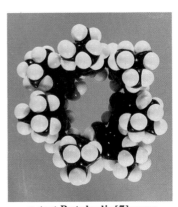

p-*tert*-Butylcalix[7]arene

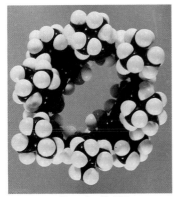

p-*tert*-Butylcalix[8]arene

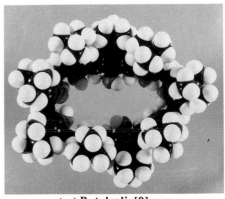

p-*tert*-Butylcalix[9]arene

Space-filling molecular models of *p-tert*-butylcalixarenes in their basket-shaped conformations, looking down into the baskets.

Contents

CHAPTER 1

From Resinous Tar to Molecular Baskets

'A man is wise with the wisdom of his time only and ignorant with its ignorance. Observe how the greatest minds yield in some degree to the superstitution of their age'

Henry David Thoreau, *Journal*, 1853

1.1   The Resinous Tar

The path of scientific research is seldom straight, often taking twists and turns quite unexpected at the outset of an odyssey. Such is the case with phenol–formaldehyde chemistry which began over a century ago in the laboratories of Adolph von Baeyer. It has developed in ways that clearly were not foreseen by this eminent scientist but which would certainly provide him with amusement and delight were he still alive to enjoy the passing scene of late twentieth century chemistry.

Johann Friedrich Wilhelm Adolph von Baeyer was one of the great organic chemists of the nineteenth century, in recognition of which he received the

Adolph von Baeyer

1

Nobel Prize in Chemistry in 1905.[1] Born in Berlin in 1835, he received his training in chemistry from Bunsen and Kekulé at Heidelberg. His first professional appointment was at the University of Ghent with Kekulé, but his career as an independent investigator really began in 1860 when, at the age of 25, he joined the staff of the Technical Institute at Berlin. Twelve years later he moved to Strasbourg as director of the chemical laboratories of the university, and here his talents as a teacher and researcher became increasingly evident to the chemical world. This led in 1875 to his appointment as successor to the great Justin Liebig at the University of Munich, a position that he held until his retirement many years later. Though Baeyer was best known for his elucidation of the structure and the synthesis of the naturally occurring dye indigo and for his work on saturated carbocyclic ring compounds, he also delved into many other areas of organic chemistry. Among these was a brief foray into the reaction of phenols with formaldehyde just at the time he was leaving Berlin for Strasbourg. Two short papers[2,3] were published from Berlin in 1872, followed by a somewhat longer one[4] from Strasbourg. These appeared in *Chemische Berichte*, and they describe the results of mixing aldehydes and phenols in the presence of strong acids.

What happens, Baeyer says in the first of these papers, is that a thickening of the reaction mixture occurs with the formation of a 'kittartige Substanz' ('cement-like substance'), and specific cases of such reactions are discussed in more detail in the second paper. For example, a mixture of benzaldehyde and benzene-1,2,3-triol (pyrogallol) was found to give a red-brown, resin-like product. Not until the third paper, however, does the field of phenol–formaldehyde chemistry actually have its beginning, the reason being that in 1872 formaldehyde was not the readily available material that it is today. Baeyer had to prepare formaldehyde by reducing CHI_3 (iodoform) with HI and red phosphorus to CH_2I_2 (methylene iodide) and then replacing the iodine atoms with oxygen moieties by treatment with silver acetate in acetic acid. This yielded a liquid which Baeyer formulated as

(*i.e.* the adduct of HCHO and CH_3CO_2H), recognizing with great insight the proclivity of the C=O group in formaldehyde to add protic nucleophiles. Thus, only with considerable effort and ingenuity did Baeyer obtain formaldehyde in sufficient quantity to allow an exploration of its reactions with phenols. Compare this with today where formaldehyde is one of the

[1] For excellent biographies of Baeyer *cf. J. Chem. Educ.*, **1930**, *7*, 1231; *ibid.*, **1929**, *6*, 1381; R. Huisgen, *Angew. Chem., Int. Ed. Engl.*, **1986**, *25*, 297.
[2] A. Baeyer, *Ber.*, **1872**, *5*, 25.
[3] A. Baeyer, *Ber.*, **1872**, *5*, 280.
[4] A. Baeyer, *Ber.*, **1872**, *5*, 1094.

world's cheapest and most readily available substances, and currently is produced in quantities greater than 2.5 million tons per year.

Using formaldehyde synthesized in this fashion, Baeyer discovered that it reacts much like the other aldehydes, producing resins when condensed with phenol in the presence of a mineral acid. However, failing to isolate pure materials from these reactions, Baeyer was unable to obtain elemental analysis data from which to propose possible structures. Only in retrospect do we see how formidable a problem he actually faced — one, in fact, that even today is not completely solved. But, it is interesting to observe how keen was his insight into the chemistry of these systems. Starting with mesitylene rather than phenol, Baeyer obtained a more tractable product from the acid-catalyzed condensation with formaldehyde which he correctly formulated as dimesitylmethane. Recognizing that hydroxymethylmesitylene should be a

mesitylene   hydroxymethylmesitylene   dimesitylmethane

logical precursor to this product, he anticipated a similar outcome from the reaction of benzene and formaldehyde. He was surprised to instead obtain a material that did not conform to this expectation but that appeared to be a 'very complex product' which, like that from phenol and formaldehyde, he was unable to characterize. The analytical tools in 1867 were primitive in comparison with today's arsenal of machines and techniques, and chemists were unable to cope with these substances which later investigations would show to be polymers. Though failing to obtain characterizable products from the phenol and formaldehyde reaction, Baeyer had, nevertheless, given birth to phenol–formaldehyde chemistry.

The next episode in this nascent field occurred in 1894 when two German chemists, L. Lederer[5] and O. Manasse,[6] independently studied the base-induced reaction between formaldehyde and phenol and succeeded in isolating *o*-hydroxymethylphenol and *p*-hydroxymethylphenol as well-defined crystalline solids. The reaction was viewed as a dehydration process in which formaldehyde reacted in its hydrated form, *viz.*

saligenin CH_2OH

[5] L. Lederer, *J. Prakt. Chemie*, **1894**, *50*, 223.
[6] O. Manasse, *Ber.*, **1894**, *27*, 2409.

o-Hydroxymethylphenol was of particular interest because it had been isolated a half century earlier as a naturally occurring compound, present in nature as the glucoside. The success of the Lederer–Manasse reaction, as it came to be called, depended on the use of mild and well-controlled conditions. Under more strenuous conditions the base-induced reaction, like its acid-catalyzed counterpart, yielded a resinous tar that resisted characterization.

As the twentieth century dawned, phenol–formaldehyde chemistry, through already a quarter of a century old, remained a rather intransigent and largely unattended area of investigation that seemed to hold little reward to explorers intrepid enough to set foot in its confines. But explorers there were — such as Blumer,[7] Storey,[8] Luft,[9] and others, all of whom tried to tame these resinous tars and find practical applications for their interesting properties. For one reason or another, however, all failed to produce materials with marketable qualitites.[10]

Success was to go to another explorer, Leo Hendrik Baekeland, who was born in 1863 in Ghent, Belgium.[11] From early childhood Baekeland's brilliance as a scholar was evident, and by 1884 at a youthful 21 he had already acquired a Doctor of Science degree. After several years of restless associations with chemistry departments at Ghent and Bruges in Belgium,

Leo Baekeland

[7] L. Blumer, British Patents 6,823 and 23,880.

[8] W. H. Storey, British Patent 8875.

[9] A. Luft, British Patent 10,218.

[10] For a brief account of the early history of phenol–formaldehyde chemistry *cf.* A. A. K. Whitehouse, E. G. K. Pritchett, and G. Barnett, 'Phenolic Resins', American Elsevier, New York, **1968**.

[11] For biographies of Baekeland *cf.* (*a*) H. V. Potter, *Chem. Ind.*, **1945**, *242*, 251; (*b*) J. Gills and R. E. Oesper, *J. Chem. Educ.*, **1964**, *41*, 224.

Cambridge and Oxford in England, and Edinburgh in Scotland, Baekeland and his young wife sailed to America in 1889, there to continue his research on photographic papers. This eventually led to a commercially successful material that he called 'Velox', a name still known to serious photographers. So successful was this product that the Velox process was purchased by George Eastman in 1900 for one million dollars, making Baekeland a wealthy man at the age of 37. Constrained by the terms of the sale from continuing to carry out photographic experiments but not wanting to abandon his passion for research, Baekeland set up a laboratory in his home, hired a number of assistants, and proceeded to explore an amazing diversity of projects. Among these was an investigation, started in 1902, of the reaction of phenol and formaldehyde. Though it had already occurred to a number of people that the hard, cement-like substance described by Baeyer might have utility as an item of commerce, only after several years of careful and painstaking work was Baekeland able to prove this premise right and show that by using a small and controlled amount of base an appealing material could be obtained. On February 18, 1907 he filed for a patent on this process[12] to make the material he called Bakelite; the age of modern synthetic plastics had begun!

The Bakelite process, ultimately described in over 400 patents issued to Leo Baekeland, constituted the first large-scale production of a synthetic plastic. Like most new things, it took time to gain acceptance; but once the induction period had been surmounted, an exponential growth phase ensued that brought great wealth to Baekeland and many others and that inspired a flood of research. One of the earliest reviews of research on the chemistry of these 'teerphenole' (tar phenols) was written by Raschig[13] in 1912 who came to the conclusion that 'Über die Chemie des Bakelits tappen wir noch vollständig im Dunkeln'. Today, three quarters of a century later, we continue to be in the dark about some of the structural details of the Bakelites. The progress that has been made, as well as the problems that remain unresolved, are recounted in a variety of books and articles.[14] As had been realized even by Baeyer in 1872, CH_2 and CH_2OCH_2 groups are the most likely types of linkages between a pair of aromatic rings in a formaldehyde–phenol condensation product. Thus, the dominant structural units in a typical resin are those shown below, two of which have acquired the trivial names 'resoles' and 'novolaks'. When resoles are heated they undergo conversion to novolak-like structures, and it is the changes that occur during this 'curing' process that have attracted most of the attention from the process engineers in the production plant as well as the research chemists in the laboratory. The former are interested in discovering how varying conditions affect the physi-

[12] L. H. Baekeland, U.S. Patent 942,699; October **1908**.

[13] F. Raschig, *Z. Angew. Chem.*, **1912**, *25*, 1939.

[14] (*a*) A Knop and L. A. Pilato, 'Phenolic Resins', Springer–Verlag, **1985**; (*b*) E. Muller in 'Methoden der Organischen Chemie' (Houben–Weyl), George Thieme, **1963**, Volume XIV/2. 'Makromolekulare Stoffe', Part 2; (*c*) N. J. L. Megson, 'Phenolic Resin Chemistry', Butterworths, London, **1958**; (*d*) R. W. Martin, 'The Chemistry of Phenolic Resins', John Wiley, New York, **1956**; (*e*) K. Hultzsch, 'Chemie der Phenolharze', Springer–Verlag, Berlin, **1950**.

resoles   novolaks   dibenzyl ethers

cal attributes of the final product; the latter are interested in discovering the chemical transformations that are occurring.

1.2   Glistening Crystals: The Zinke Products

It is the outcome of a study of the 'curing' phase of the phenol–formaldehyde process that introduces the next episode in this story and brings into existence the central subject of this book. In 1942 Alois Zinke, a professor of chemistry at the University of Graz in Austria, and his coworker Erich Ziegler decided to 'simplify' the problem by looking not at phenol but at *p*-substituted phenols in the condensation reaction with formaldehyde.[15] Phenol itself reacts at the *ortho* and *para* positions to form highly cross-linked polymers in which each phenolic residue is attached to three other phenolic residues, *viz.*

A *para*-alkylphenol, on the other hand, can react only at the two *ortho*-positions, thereby reducing the cross linking possibilities to the formation of a linear polymer, *viz.*

---

[15] For an account of the contributions of v. Euler, Hultzsch, and Zinke to phenol–formaldehyde chemistry *cf.* ref. 14*c*.

As a consequence, it was hoped that more tractable products would be obtained and that insight into the curing process might thereby be gained. A typical experiment in this study by Zinke and Ziegler[16] is described in 1944 as follows: 'When 50 g of the resin, which is obtained by heating 100 g of *p-tert*-butylphenol, 100 ml of 3 N NaOH, and 97 g of a 35% formaldehyde solution, is heated with stirring with 200 g of linseed oil, it dissolves at 100–120 °C. At 140–160 °C there is vigorous foaming and turbidity appears which increases greatly upon further heating at 200–220 °C. The resulting brown waxy paste is stirred with ethyl acetate, washed thoroughly, and reprecipitated from CCl_4 or $CHCl_3$ with alcohol to give a crystalline product as platelets or rosettes that decompose above 300 °C.' A product of presumably identical composition had been described in less detail three years earlier by these same authors[17] at which time an elemental analysis commensurate with a $C_{11}H_{14}O$ structure was obtained. The lack of reactivity of the compound with HBr signaled the absence of ether linkages, but no structure was drawn in the 1941 paper. By 1944, however, the idea of *cyclic structures* had sprung to several minds and was 'in the air'. Joseph Niederl and his coworker Heinz Vogel[18] had proposed a cyclic tetrameric structure for compounds obtained by the acid-catalyzed treatment of aldehydes and phenols, as described in the next section, so it was a propitious time for Zinke to postulate a cyclic tetrameric structure (**1**) for his product as well.[19]

1

Alois Zinke[20] was born in Bärnbach, a region of Voitsberg, Austria in 1892. His early schooling was acquired in Voitsberg, but in 1902 he went to Graz for the completion of his training, acquiring a PhD from the University of Graz in 1915 under the tuteledge of Roland Scholl, and the remainder of

[16] A. Zinke and E. Ziegler, *Ber.*, **1944**, *77*, 264.
[17] A. Zinke and E. Ziegler, *Ber.*, **1941**, *B74*, 1729.
[18] J. B. Niederl and H. J. Vogel, *J. Am. Chem. Soc.*, **1940**, *62*, 2512.
[19] In a later paper (see ref. 22) Zinke refers to an 'observation by H. Hönels' in connection with the cyclic tetrameric structure, but nothing beyond this cryptic allusion is offered to indicate the nature of Hönel's contribution.
[20] For a biography of A. Zinke *cf.* E. Ziegler, *Scientia Pharmaceutica*, **1951**, 209; *idem, Arznei-mittel-Forschung*, **1963**, *12*, 208. We are deeply indebted to Prof. Dr. Helge Wittmann for making copies of these biographies of her father available to us.

his life was spent in that city. He started his professional career at the University of Graz, transferred his allegiance to the Technische Hochschule in Graz for a brief four years, and then returned to the university as professor of pharmaceutical chemistry. In 1941 he was appointed director of the Institutes für Organische und Pharmazeutische Chemie. During his distinguished career, which extended to his death in 1963, he carried out research in several areas, but he is best known for his work with phenol–formaldehyde resins. When he started this work in 1936, there already were several well-established investigators in the field, including Koebner, Megson, v. Euler, Hultzsch, and Adler. Undaunted by his competition, Zinke initiated his own program and went on to make numerous and important contributions, including the seminal discovery of the cyclic oligomers.

Alois Zinke

The product that Zinke and Ziegler obtained in their first experiment[17] yielded a crystalline acetate (mp 314 °C) which had the surprisingly high molecular weight of 1725. On the premise that the compound from which the acetate is derived is a cyclic oligomer, this value indicates the presence of 8 (or possibly 9) *p-tert*-butylphenol units in the cyclic array. However, this seemed so unlikely that they desisted from drawing any structure in the 1941 paper and waited until 1944 before proposing the intuitively more appealing cyclic tetramer structure.

Zinke's choice of a cyclic tetrameric structure for the product obtained from the base-induced condensation of *p-tert*-butylphenol and formaldehyde seemed entirely logical, and the high molecular weight value for the acetate was dismissed as a complication caused by mixed crystal or molecular compound formation. Additional examples of the reaction, which Zinke refers to as one involving 'nicht alkalifrei gewaschenen Resols in Leinöl' ('resols, not washed entirely free of base, in linseed oil') were reported in a short paper four years later[21] in which *p*-phenylphenol, *p*-cyclohexylphenol, and *p*-benzylphenol are described as giving high melting, organic solvent-insoluble materials. Cyclic tetrameric structures were assigned to all of these products, although no experimental data were provided. Not until 1952 was further information forthcoming in a paper that is the most detailed of all of the Zinke publications on the cyclic oligomers. Published with R. Kretz, E. Leggewie, and K. Hössinger,[22] it provided additional examples of phenols reacting with formaldehyde to give high melting, organic solvent-insoluble substances. More importantly, it provided solid evidence in support of the cyclic tetrameric structure. Zinke recognized that an unequivocal proof of structure had not been given in his previous publications, and he implicitly acknowledged that the molecular weight of the acetate of the product from *p*-tert-butylphenol and formaldehyde constituted a puzzling complication. Thus, it must have been a great relief to find that the acetate prepared from *p*-(1,1,3,3-tetramethylbutyl)phenol and formaldehyde, isolated as needles with mp 333 °C, had a cryoscopic molecular weight of only 876. The molecular weight calculated for the cyclic tetramer from this phenol is 873, which is in good agreement with the experimental value. Satisfied with this piece of evidence, Zinke concluded that all of the products that he had isolated were pure, cyclic tetramers. Through these pioneering efforts Zinke and his coworkers introduced to the chemical world a series of compounds whose structures appeared to be accurately and adequately described as the cyclic tetramers (**2**), prepared from *p*-methyl-, *p*-tert-butyl, *p*-tert-pentyl-, *p*-(1,1,3,3-tetramethylbutyl)-, *p*-cyclohexyl-, *p*-phenyl-, and *p*-benzylphenol.

The four papers by Zinke that are discussed above represent only a small fraction of his total scientific output, but in the context of the present book they are his most important. They gave glimpses of the treasures that are to be found in the resinous tar of a phenol–formaldehyde condensate, and they provided an important part of the scaffolding on which calixarene chemistry has been built. No matter how original one's contributions may appear to be,

[21] A. Zinke, G. Zigeuner, K. Hössinger, and G. Hoffmann, *Monatsh*, **1948**, *79*, 438.
[22] A. Zinke, R. Kretz, E. Leggewie, and K. Hössinger, *Monatsh.*, **1952**, *83*, 1213.

however, antecedents can usually be found lurking in the shadows. The case of the cyclic tetramer is no exception, for Raschig had suggested such a structure back in 1912 in the paper that was cited earlier in this chapter.[13] However, his idea must be regarded as a lucky guess rather than a rational choice based on experimental evidence, and to Zinke should go the honor of true parentage. But, as we shall see in later sections of this chapter, the complete character of his progeny was yet to be revealed.

1.3   More Crystals: The Niederl Products

At this point we must retrace our steps, look once again at Baeyer's experiments in 1872, and follow another trail of events that flowed from these beginnings. Resorcinol was among the reactants that Baeyer used in his investigation with phenols and aldehydes, and he discovered that it reacts with aldehydes such as benzaldehyde and acetaldehyde under acidic conditions to produce crystalline, high melting compounds. However, the products lacked the dye-stuff properties that Baeyer was seeking, so he did not pursue their characterization and simply concluded that they were 1:1 condensation products. The reaction was reinvestigated a decade later by Michael[23] who succeeded in isolating a pair of crystalline materials for which he postulated cyclic dimeric structures. Similar experiments were carried out in 1894 by Möhlau and Koch[24] and again in 1904 by Liebermann and Lindenbaum,[25] but the compounds then remained unnoticed until 1940 when Neiderl and Vogel[18] reinterpreted their chemistry. These workers isolated solid, high melting condensation products from the reaction of resorcinol with acetaldehyde, propionaldehyde, and isovaleraldehyde and concluded, primarily on the basis of molecular weight determinations, that the products were best represented as cyclic tetramers **3**. In contrast to the Zinke tetrols in which four OH groups are intraannular, the eight OH groups in the Niederl octols are extraannular.[26] Extending their studies to include

3

[23] A. Michael, *Am. Chem. J.*, **1883**, *5*, 338.

[24] R. Möhlau and P. Koch, *Ber.*, **1894**, *27*, 2887.

[25] C. Liebermann and S. Lindenbaum, *Ber.*, **1904**, *37*, 1171.

[26] Appreciation is expressed to Dr. Högberg for providing a copy of his excellent thesis (Royal Institute of Technology, Stockholm, Sweden, **1977**) in which the 'intraannular' and 'extra-annular' terminology is suggested.

phenol–formaldehyde products, Niederl and McCoy[27] repeated an earlier experiment of Koebner[28] who had claimed that a linear trimer is produced in the acid-catalyzed reaction of *p*-cresol with its bis-hydroxymethyl derivative. However, clearly influenced by their work on the resorcinol–aldehyde products and relying on what they believed to be their more accurate molecular weight determinations, Niederl and McCoy[27] postulated that the Koebner product was actually the cyclic tetramer (2, R = Me). That they were wrong and that Koebner's linear trimer structure was correct was shown first

Koebner product

in 1950 by Finn and Lewis,[29] and then again in 1961 by Foster and Hein.[30] Undoubtedly, the controversy arising from the incorrect assignment of the structure of the Koebner product cast doubt for some time on Neiderl's claim of a cyclic tetrameric structure for the resorcinol–aldehyde products. Subsequent definitive experiments, however, showed the latter to be cyclic, as Neiderl had claimed, so the honor of being the first to assign correctly cyclic tetrameric structures to phenol–aldehyde products on the basis of experimental evidence perhaps should really go to Niederl. It seems likely that his work had an influence on Zinke's assignment of structure.

1.4   Cyclic Tetramers: Proofs of Structure

By the 1950's the work of Zinke on the cyclic tetramers had become known to chemists interested in phenol–formaldehyde chemistry. Among these were B. T. Hayes and R. F. Hunter in the Research & Development Department of Bakelite Ltd, Tyseley, Birmingham, England. In 1956 this pair of chemists published a short account[31] of what they termed 'a rational synthesis of cyclic tetranuclear *p*-cresol novolak', following this in 1958 with a somewhat longer and more detailed account.[32]

The Hayes and Hunter synthesis, outlined in Figure 1.1, provides a classic example of the use of a blocking group which is added at one point in a process to protect a reactive site and then removed at a later point to reopen that site to reaction. The protecting group that they chose was the bromine atom, introduced in the opening step of the sequence into one of the *ortho* positions of *p*-cresol to give 2-bromo-4-methylphenol (5). Base-induced hydroxy-

[27] J. B. Niederl and J. S. McCoy, *J. Am. Chem. Soc.*, **1943**, *65*, 629.
[28] M. Koebner, *Z. Angew. Chem.*, **1933**, *46*, 251.
[29] S. R. Finn and G. J. Lewis, *J. Soc. Chem. Ind.*, **1950**, *69*, 132.
[30] H. M. Foster and D. W. Hein, *J. Org. Chem.*, **1961**, *26*, 2539.
[31] B. T. Hayes and R. F. Hunter, *Chem. Ind.*, **1956**, 193.
[32] B. T. Hayes and R. F. Hunter, *J. Appl. Chem.*, **1958**, *8*, 743.

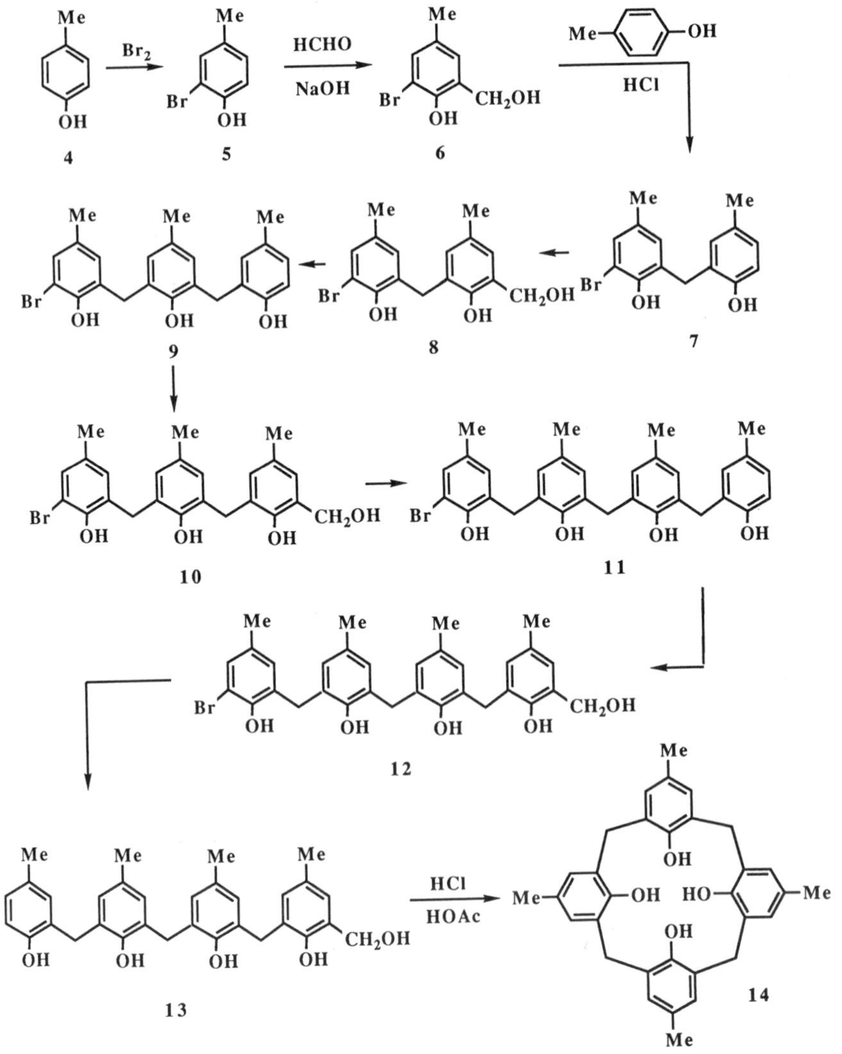

Figure 1.1 *Hayes and Hunter stepwise synthesis of a cyclic tetramer*

methylation of **5** yielded 2-bromo-4-methyl-6-hydroxymethylphenol (**6**) which was then treated with concentrated HCl and a large excess of *p*-cresol, heated at 70 °C for 18 hours, and worked up to give a reasonably good yield of 3-bromo-2:2'-dihydroxy-5:5'-dimethyldiphenylmethane (**7**). Two repetitions of this *pas de deux, viz.* base-induced hydroxymethylation followed by acid-catalyzed arylation, produced **8** and **9** and then **10** and **11**. One more hydroxymethylation yielded **12** from which the bromine was removed by catalytic hydrogenation to afford the penultimate product **13**. Acid-catalyzed treatment of **13** under high dilution conditions effected cyclization to **14** in unstated yield. The material that was obtained was described as a light brown solid that did not melt below 300 °C, that was soluble in a variety of organic solvents and that did not undergo coupling with benzenediazonium chloride (indicating the absence of any reactive *ortho* or *para* positions). A molecular weight determination gave a value of 525, in reasonably close agreement with the calculated value of 480. Employing the then rather new technique of infrared analysis, Hayes and Hunter observed an absorption band at 854 cm^{-1} of the tetraacetate of **14** indicative of a 1,2,4,6-substitution pattern on the aromatic rings. Elemental analyses of **14** as well as its tetraacetate gave results commensurate with a cyclic tetramer, though in both cases the inclusion of a certain amount of water in the product had to be invoked to bring the observed values within range of the calculated values.

Hayes and Hunter concluded that their synthesis showed that 'cyclic structures of the type **14** may be produced under suitable environmental conditions in the hardening process in phenol–formaldehyde resins' and that it confirms 'that the analogous novolaks which Zinke and his collaborators claimed to have obtained by heating resoles such as 2:6-dimethylol-4-*tert*-butylphenol are, at least, sterically possible'. That it did, indeed, establish the validity of a cyclic tetrameric structure is vouchsafed, and this synthesis, though simple and straightforward in concept, represents a significant contribution to the literature of phenol–formaldehyde chemistry. It was tacitly accepted as a proof of structure for all of the Zinke products in spite of the fact that no direct comparison was made between **14** and the compound obtained by a Zinke reaction from *p*-cresol. In fact, even today, no direct comparison of **14** with a product obtained *via* the Zinke reaction has ever been made, although an indirect connection has been established in a fashion that is discussed in a later chapter. The Hayes and Hunter synthesis represents a highly laudable achievement, and it established the basis for an extensive program carried out some years later by Hermann Kämmerer and his coworkers. However, it may have delayed the more careful investigation of the Zinke reaction because of an implied assumption that it provided the capstone in the structure proof of the Zinke products.

Shortly after Zinke's 1952 paper describing the application of his procedure to a variety of *p*-substituted phenols, but before the Hayes and Hunter papers had appeared in print, another participant entered the field. Whereas Zinke as well as Hayes and Hunter were interested in the cyclic tetramers as a facet of the phenol–formaldehyde problem, the new entrant was interested in the cyclic structures *per se*. John W. Cornforth, a British chemist who was destined to win a Nobel Prize two decades later for his work on the stereochemistry of enzyme-catalyzed reactions, was interested in 1955 in the preparation of tuberculostatic substances. Among the compounds tested for this purpose were a variety of oxyethylated phenols, including linear phenol–formaldehyde oligomers as well as the then recently described Zinke products. When Cornforth and his coworkers[33,34] repeated the Zinke

[33] J. W. Cornforth, P. D'Arcy Hart, G. A. Nicholls, R. J. W. Rees, and J. A. Stock, *Br. J. Pharmacol.*, **1955**, *10*, 73.

[34] J. W. Cornforth, E. D. Morgan, K. T. Potts, and R. J. W. Rees, *Tetrahedron*, **1973**, *29*, 1659.

reaction starting with *p-tert*-butylphenol, two materials rather than a single product were isolated. Both were crystalline, sparingly soluble compounds with high but non-identical melting points. They had elemental analyses compatible with a $(C_{11}H_{14}O)_n$ formula, and both possessed the physical chemical properties characteristic of a cyclic oligomer. When *p*-(1,1,3,3-tetramethylbutyl)-phenol (often referred to as *p-tert*-octylphenol) was employed as starting material the outcome was similar, *viz.* two compounds with essentially identical properties but with somewhat different melting points were isolated. For convenience these substances were referred to by Cornforth as the high melting compounds HOC and HBC from *p-tert*-octylphenol and *p-tert*-butylphenol, respectively, and the low melting compounds LOC and LBC from *p-tert*-octylphenol and *p-tert*-butylphenol, respectively.

John Cornforth

Since Zinke's work had given no indication that the products from the base-induced reactions of *p*-substituted phenols and formaldehyde might be mixtures, the isolation of a pair of compounds from both the *p-tert*-butylphenol and the *p-tert*-octylphenol reactions was disturbing. One of the obvious structural possibilities for 'the other' compound is that it is a linear oligomer. However, this was quickly ruled out by the elemental analyses which indicated the presence of the same ratio of CH_2 groups (from HCHO) and phenolic residues in all of the compounds isolated. Also, the failure of either the high-melting or low-melting compounds to react with *p*-nitrobenzene-diazonium chloride indicated the absence of any reactive positions on the aromatic rings. The possibility that the high-melting and low-melting compounds might be cyclic oligomers of different ring sizes seemed to be negated by the results of *X*-ray crystallography. Dorothy Crowfoot Hodgkin, one of Britain's most eminent *X*-ray crystallographers, reported the following: 'Both HOC and HBC have very complex crystal structures in which the asymmetric units have respectively four and three times the weight required for a tetrapolymer. Formally they admit several solutions for the molecular complexity of these compounds. HOC acetate is rather simpler; the molecule here from the *X*-ray evidence is most probably a 4-polymer (space group P1), but might be either an 8-polymer, or possibly a 7-polymer (P1), since difficulty was experienced in measuring accurately the lattice constants of the triclinic crystals. LOC and LBC both have crystal structures which indicate they are tetrapolymers. Formally, in the case of LOC, the molecule might correspond with a twofold polymer or with an eightfold polymer having a center of symmetry in the molecule. The latter is, however, not a stereochemically probable solution, and the molecule here almost certainly corresponds with the crystal asymmetric unit. LBC crystallizes in the tetragonal system; here, the molecule is proved not only to be a fourfold polymer but also to have either a fourfold or a fourfold alternating axis of symmetry'.

With this evidence in hand, Cornforth stated that 'Though they do not establish with certainty the molecular complexity of all the compounds the crystallographic data are consistent with the view that all four condensation products have a cyclic tetrameric structure'. He goes on to say that cryoscopic molecular weight determinations of HBC, LBC, HOC, and LOC all reinforce this conclusion and 'refute the possibility, admitted by the crystallographic data, of molecules containing eight phenolic nuclei'. But, if both the high-melting and low-melting compounds are cyclic tetramers how does one account for the differences in their properties? To answer this question Cornforth proposed that the compounds are diastereoisomers arising from hindered rotation. Examination of molecular models revealed the possibility of four different structures, shown in Figure 1.2, and Cornforth assumed that 'the phenolic nuclei cannot rotate about the bonds joining them to the methylene groups'. Thus, each of these four structures, today called conformers, should be capable of an independent existence. Since they bear a

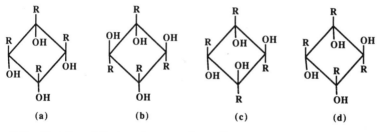

Figure 1.2 *Rotational diastereoisomers of a cyclic tetramer*

diastereoisomeric relationship to one another, they would be expected to have different chemical and physical properties. Other investigators came to similar conclusions from an examination of models. For example, Ballard, Kay, and Kropa[35] constructed these four conformers from space filling molecular models, inspection of which led to their statement that the 'phenolic nuclei cannot rotate around the methylene linkage'.

By the late 1950's, therefore, the evidence seemed to be quite conclusive that the Zinke reaction produces only cyclic tetramers. Although Cornforth's experiments showed that the reaction was not as clean as seemed to be implied by Zinke's descriptions, the notion of the cyclic tetrameric structure appeared to be firmly supported and accepted. Only later would the truth of Henry David Thoreau's insight be reaffirmed, the problem in this case being too great a reliance on the accuracy of molecular models in reflecting the magnitude of rotational barriers.

1.5   The Petrolite Chapter

The Petrolite Corporation is located in Webster Groves, Missouri, about four miles outside the city limit of St Louis and about ten miles from the Mississippi River. Started in 1916 by a young pharmacist named William S. Barnickel, it grew from very modest beginnings to become one of St Louis' larger companies. Its original product line was demulsifiers for resolving crude oil emulsions, and still today a significant share of its total revenue comes from the sale of such materials. It is one of these that plays a critical role in the present story.

Crude oil as it comes out of the ground is usually mixed with water, generally as an emulsion that is difficult to break. What gave Barnickel his big chance in business and allowed him to turn his back on pharmacy was his discovery, after several frustrating years of testing by trial and error, that ferrous sulfate was effective in breaking the emulsified oil from the large Caddo oil field near Schreveport, Louisiana. It soon become apparent, though, that ferrous sulfate was far from being a universal oil demulsifier. In fact, it was almost unique to that emulsion, and it turns out that oil from wells in various locations around the world all have their individual characteristics. A demulsifier that works on one will not necessarily work on another. Thus, for Barnickel's newly emerging Petrolite Company to offer products for the world's spectrum of oil wells it was necessary that a range of tailor-made substances be made available. To provide this range of products a scientist named Melvin DeGroote was hired in 1924. By the time that this remarkable man retired in 1960 he had been granted almost a thousand patents on crude oil demulsifiers, making him the world's record holder for the greatest number of U.S. chemical patents. Among the numerous products that DeGroote and his burgeoning research staff discovered were the oxyalkylated alkylphenol–formaldehyde resins. For example, when the product obtained

---

[35] J. L. Ballard, W. B. Kay, and E. L. Kropa, *J. Paint Technology,* **1966,** *38,* 251.

from the acid-catalyzed condensation of *p*-nonylphenol and formaldehyde was treated with ethylene oxide it produced a surfactant that had excellent demulsifying capabilities. Then, it was discovered that even better demulsifiers resulted when base-induced condensation of the alkylphenol and formaldehyde was employed, and in the 1950's a Petrolite demulsifier went on the market that was made by oxyalkylating the product from *p-tert*-butylphenol and formaldehyde, the assumption being that it was a linear oligomer with the structure shown below.

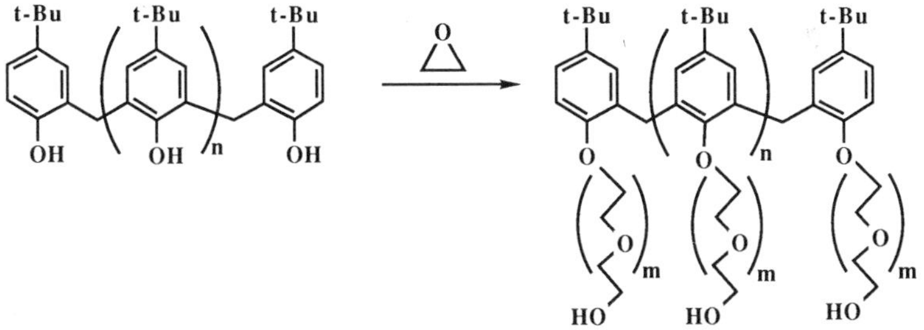

This material was sold as a solution in a mixture of aromatic hydrocarbons. From the outset, however, complaints were received from the customers in the oil fields as well as from the workers in the Petrolite production plant, the problem being that sludges precipitated from the solution, making the handling and application of the product difficult. Not knowing the cause of the problem, the plant engineers sought the help of the chemists in the research laboratory. A five person team consisting of Franklin Mange, Rudolf Buriks,

Melvin DeGroote (in picture)

Franklin Mange   Rudolf Buriks   Alan Fauke   Jack Ludwig   John Munch

Petrolite chemists

Alan Fauke, John Munch, and Jack Ludwig addressed the problem by devising a laboratory procedure that simulated the one used in the production plant. Much of this work was done by John Munch who prepared a slurry of *p-tert*-butylphenol and paraformaldehyde in xylene, added a small amount of 50% KOH solution, and refluxed the mixture for several hours in an apparatus equipped with a Dean and Stark trap to remove water from the reaction mixture. During the course of the reaction a copious precipitate formed which was removed by filtration and found to be a very high melting, very insoluble compound crystallizable from chloroform as very small, glistening needles. Intrigued by these properties, the Petrolite scientists proceeded to search the chemical literature, and they discovered the existence of the chemistry that has been recounted in the first two sections of this chapter. On the basis of this information, they concluded that although their recipe was different from the one described by Zinke, the material that they had isolated must be a Zinke cyclic tetramer. Patents were eventually filed[36,37,38] by the Petrolite group in 1976–7 describing what has since come to be known as the 'Petrolite Procedure' for making cyclic oligomers. These patents were thought to represent more correctly the composition of the demulsifier than the earlier ones in which linear oligomeric structures were assigned.

1.6   Cyclic Tetramers and the Quest for Enzyme Mimics

In 1947 David Gutsche, fresh from graduate school at the University of Wisconsin where he received a PhD degree under the direction of William S. Johnson, joined the Department of Chemistry of Washington University which is located on the outskirts of St Louis only a few miles from the Petrolite Corporation. One of the results of this geographic proximity was a consulting arrangement that began in 1949 and continues still today. Out of this association came an awareness on Gutsche's part of phenol–formaldehyde chemistry in general and cyclic oligomeric chemistry in particular. Consequently, when Gutsche became intrigued in the early 1970's with the newly emerging area of bioorganic chemistry known as enzyme mimics, the Zinke cyclic tetramers sprang to mind as potential candidates for molecular baskets. The basic idea of enzyme mimic building is to construct a receptor for a substrate molecule and equip the receptor with the functional groups that are appropriate for interacting in some fashion with the substrate molecule (see diagram opposite). To pursue this possibility a research program was initiated in 1972 in the laboratories at Washington University with the goal of exploring the Zinke compounds as cavity-containing substances appropriate for enzyme mimic building.

[36] R. S. Buriks, A. R. Fauke, and J. H. Munch, U.S. Patent 4,259,464; filed 1976, issued 1981.
[37] R. S. Buriks, A. R. Fauke, and F. E. Mange, U.S. Patent 4,098,717; filed 1977, issued 1978.
[38] R. S. Buriks, A. R. Fauke, and F. E. Mange, U.S. Patent 4,032,514; filed 1976, issued 1977.

David Gutsche

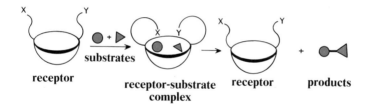

The choice of the Zinke cyclic tetramers as potential enzyme mimics seemed to Gutsche to be a rational one. He considered the recently-discovered crown ethers to be unappealing, because in their simplest form they are loops rather than cavities. The subsequent work of Cram,[39] and Lehn[40] and their respective coworkers that led to a Nobel Prize in 1987, of course, proved how useful the crown ethers can actually be when appropriately modified. The cyclodextrins were dismissed because they are natural rather than synthetic products and also because their potential as enzyme mimics was already being vigorously investigated by several superb scientists.[41] In contrast to the simple crown ethers, the Zinke compounds are baskets rather than loops. In contrast to the cyclodextrins, they are accessible by simple laboratory synthesis. And in contrast to both the crown ethers and

[39] D. J. Cram, *Angew. Chem., Int. Ed. Engl.*, **1986**, *25*, 1039.
[40] J.-M. Lehn, *Angew. Chem., Int. Ed. Engl.*, **1988**, *27*, 89.
[41] For recent reviews *cf.*: (*a*) M. L. Bender and V. T. D'Souza, *Acc. Chem. Res.*, **1987**, *20*, 146; (*b*) R. Breslow, *Adv. Enzymol. Relat. Areas. Mol. Biol.*, **1986**, *58*, 1–47; (*c*) I. Tabushi, *Acc. Chem. Res.*, **1982**, *15*, 66.

the cyclodextrins, they appeared to be known to only a handful of chemists in 1972 and were largely unexplored. The information on these compounds that had accumulated by the early 1970's, as recounted in the first two sections of this chapter, suggested them as the ideal, synthetic cavity-containing compounds, *viz.* easily accessible by a one-step synthesis that could be carried out on a variety of *p*-substituted phenols. So attractive a class of compounds called for an engaging name.

Zinke[21] referred to his cyclic tetramers as 'Mehrkernmethylenephenol-verbindungen', an aptly descriptive teutonic name; Hayes and Hunter[32] called them 'cyclic tetranuclear novolaks', employing a term descriptive of phenol–formaldehyde oligomers lacking hydroxymethyl groups; Cornforth[33] sought a more systematic nomenclature and called them '1:8:15:22-tetra-hydroxy-4:11:18:25-tetra-*m*-benzylenes'; *Chemical Abstracts*[42] names the basic ring structure of the cyclic tetramer as '[19.3.1.1^{3,7}1^{9,13}1^{15,19}]octacosa-1(25),3,5,7(28),9,11,13(27),15,17,19(26)21,23-dodecaene'. In a now widely used nomenclature invented in 1951 by Cram and Steinberg[43,44] these compounds are classed as [1$_n$]metacyclophanes. However, something more pictorially and descriptively appealing seemed .appropriate for the Zinke cyclooligomers.

The nonplanar character of the cyclic tetramers was pointed out by Cornforth[33] as well as by Megson[45] and by Ott and Zinke.[46] Perceiving a similarity between the shape of a Greek vase called a *calix crater*, shown on the right-hand side of Figure 1.3 and a space-filling molecular model of the form of the Zinke cyclic tetramer in which all of the aryl moieties are oriented in the same direction, as shown on the left-hand side of Figure 1.3, Gutsche coined the name 'calixarene' in 1975 (although it did not appear in print until 1978).[47] The name is derived from the Greek *calix* meaning 'vase' or 'chalice', and *arene* which indicates the presence of aryl residues in the macrocyclic array. Although the name has not yet made its way into the lexicon of IUPAC nomenclature, it appears to have been generally accepted by the chemists working with these compounds.

The first set of experiments in the Washington University program dealt with the scope and limitations of the phenol–formaldehyde cyclization reaction to determine which phenols react to yield high melting products and which ones do not. The results of this investigation suggested that a variety of *p*-substituted phenols yield the desired product, including *p*-methyl-, *p*-tert-butyl-, *p*-phenyl-, *p*-methoxy-, and *p*-carbomethoxyphenol. If a high melting, organic solvent-insoluble precipitate was produced it was assumed to be the

[42] A. M. Patterson, L. T. Capell, and D. F. Walker, The Ring Index, 2nd ed., American Chemical Society, Washington, D.C., 1960, Ring Index No. 6485. We are indebted to Dr. K. L. Loening of Chemical Abstracts Services for helpful guidance in nomenclature.
[43] D. J. Cram and H. Steinberg, *J. Am. Chem. Soc.*, **1951**, *73*, 5691.
[44] IUPAC Tentative Rules for Nomenclature of Organic Chemistry, Section E. Fundamental Stereochemistry; *cf. J. Org. Chem.*, **1970**, *35*, 284.
[45] N. R. L. Megson, *Oesterr. Chem. Ztg.*, **1953**, *54*, 317.
[46] R. Ott and A. Zinke, *Oesterr. Chem. Ztg.*, **1954**, *55*, 156.
[47] C. D. Gutsche and R. Muthukrishnan, *J. Org. Chem.*, **1978**, *43*, 4905.

Figure 1.3 *Space-filling molecular model of a cyclic tetramer (left) and a* calix crater *(right)*

cyclic tetramer, since these structures were considered at the time to have been well-established by precedent. These results were confidently reported in 1975 by Gutsche, Kung, and Hsu,[48] first in Honolulu at an NSF-sponsored East–West Cultural Exchange Symposium and later at the Midwest Regional Meeting of the American Chemical Society in Carbondale, Illinois. At this juncture, the pathway to producing a wide variety of appropriately substituted molecular baskets seemed assured. Of particular interest in this respect was the product from *p*-phenylphenol, because space-filling models show that it has a very deep cavity that should be capable of forming complexes with molecules of appreciable size. Two years after these initial reports a paper by Patrick and Egan appeared in the *Journal of Organic Chemistry*[49] that essentially duplicated these experiments. The authors of this study, using a slightly modified Petrolite Procedure involving potassium *tert*-butoxide as the base and tetralin as the solvent, reported the obtention of cyclic tetramers in precisely the same five cases that Gutsche had studied. Since the senior author Timothy Patrick was acquainted with the research at Washington University, the appearance of this paper was greeted less than enthusiastically in St Louis. The end result was beneficial, however, for it provoked the Gutsche group into taking a closer look at the Petrolite Procedure, the Zinke Procedure, and the procedure employed by Patrick and Egan, leading eventually to a more complete understanding of the cyclooligomerization process.

[48] C. D. Gutsche, T. C. Kung, and M.-L. Hsu, Abstracts of 11th Midwest Regional Meeting of the American Chemical Society, Carbondale, IL, 1975, No. 517.
[49] T. B. Patrick and P. A. Egan, *J. Org. Chem.*, **1977**, *42*, 382; *idem, ibid.*, **1977**, *42*, 4280.

1.7   Unraveling the Literature

A comparison of the Petrolite Procedure and the Patrick and Egan modification using the five *p*-substituted phenols listed above showed considerable differences between the products in their melting points. Further examination of the IR spectra of the products revealed small but persistent differences (*e.g.* well-resolved bands in some preparations at 800 and 780 cm^{-1} but in others only a shoulder at 800 cm^{-1} along with a well-resolved band at 780 cm^{-1}). Silylation followed by thin layer chromatography indicated the presence of more than one compound in every instance. That the trimethylsilyl derivatives were not simply conformational isomers was demonstrated by hydrolytic removal of silyl groups which yielded two, or more, different parent compounds. With the realization that mixtures comprising compounds of *different gross structures* were being produced, a careful and detailed study of the reaction with *p-tert*-butylphenol was undertaken.

Before discussing this investigation, however, let us look at still another set of experiments that impeded the correct interpretation of calixarene chemistry for several years. In 1972 Hermann Kämmerer[50] at the University of Mainz undertook a reinvestigation of the Hayes and Hunter method for the stepwise synthesis of calixarenes, improving and extending this synthesis by preparing a variety of calixarenes carrying methyl and/or *tert*-butyl groups in the *p*-positions. Among the analytical techniques applied to these materials was that of temperature-dependent ^{1}H NMR which revealed that the cyclic tetramers are considerably more flexible than had been thought by Cornforth and others. Space-filling molecular models, which earlier investigators had invoked to support the idea of severely restricted rotation, upon reinspection in the light of these NMR data are discovered to be 'softer' than had been supposed. Kämmerer found, for example, that the ^{1}H NMR spectrum of a cyclic tetramer at 20 °C shows a pair of doublets arising from the protons of the CH$_2$ bridges between the aryl rings, while a spectrum at 60 °C shows only a singlet.[50] This change in the character of the resonances from the CH$_2$ protons (discussed in Chapter 4) was initially interpreted by Kämmerer in terms of the Cornforth isomers, form-**a** in Figure 1.2 converting to form-**b** upon heating. Three years later,[51] however, he reinterpreted the data in the currently accepted terms of a mirror image conformational interconversion of form-**a** structures. Meanwhile, in the early 1970's Munch[52] independently made similar observations for the ^{1}H NMR spectra of the Petrolite product. Unfortunately, the paper detailing his results was not accepted for publication when originally submitted, and it appeared in print after the Kämmerer 1975 paper. The studies of both Kämmerer and Munch indicated the presence of a flexible system that undergoes conformational inversion at the rate of *ca.* 150 sec^{-1} at room temperature, calling into question the validity of Cornforth's postulate that his LBC–HBC and LOC–HOC pairs of

[50] H. Kämmerer, G. Happel, and F. Caesar, *Makromol. Chem.*, **1972**, *162*, 179.
[51] G. Happel, B. Mathiasch, and H. Kämmerer, *Makromol. Chem.*, **1975**, *176*, 3317.
[52] J. H. Munch, *Makromol. Chem.*, **1977**, *178*, 69.

compounds are *conformational* isomers. Also, the identity of the rates of inversion of the Petrolite product and the Kämmerer product (known unequivocally to be the cyclic tetramer), appeared to provide additional proof for the cyclic tetrameric character of the Petrolite product. As the next experiments show, however, the agreement is coincidental — one of Nature's devious accidents.

Returning to the research at Washington University, we now consider the results of the study of the product from the Petrolite Procedure using *p-tert*-butylphenol and formaldehyde. The crude product is described[47] as a colorless substance that melts at *ca.* 360—375 °C. Two recrystallizations from chloroform raises the melting point to above 400 °C and produces colorless, fine needles that display very simple ^{1}H NMR and ^{13}C NMR spectra. The ^{13}C NMR spectrum in particular provides a telling comparison between the linear and cyclooligomers, as illustrated in Figure 1.4. An osmometric molec-

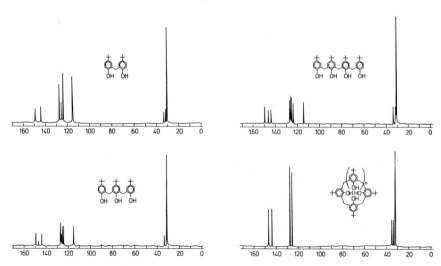

Figure 1.4 *^{13}C NMR spectra of linear oligomers and the cyclic tetramer from* p-tert-butylphenol

ular weight determination gives a value of 1330 in agreement with a cyclic octamer, but the strong mass spectral signal at *m/e* 648 and the similarity of the dynamic ^{1}H NMR spectrum with that of authentic cyclic tetramer gave substance to the thought that this high value arises from association of a pair of cyclic tetramers. However, small signals at *m/e* values higher than 648 consistently appeared in the mass spectrum, suggesting that the *m/e* 648 signal might *not* arise from the mono-cation of the parent molecule. This suspicion was confirmed when a trimethylsilyl derivative showed a strong mass spectral signal at *m/e* 1872, and with this piece of evidence it finally became apparent that the compound is *not* the cyclic tetramer but is the *cyclic octamer* (**15**). The validity of this conclusion was later convincingly cor-

15

roborated by *X*-ray crystallography.[53] The recognition of this structure as the cyclic octamer resolved some of the apparent discrepancies in the literature that had so long plagued calixarene chemistry.

As discussed above, Zinke's first paper[17] reported a product from *p-tert*-butylphenol and formaldehyde that yielded an acetate showing a molecular weight of 1725, close to that expected for the acetate of *p-tert*-butyl cyclic octamer. It seems quite likely that this, indeed, is what he isolated from this particular reaction mixture, although there is no doubt that some of the products reported in Zinke's later papers are cyclic tetramers. Also, it should be recalled that the *X*-ray crystallographic data on the high-melting compound isolated by Cornforth from *p-tert*-butylphenol or *p-tert*-octylphenol[33] admitted the possibility of a cyclic octameric structure, and it is now known that the octamer was, in fact, the product he had isolated. What the Petrolite chemists had patented[36–38] and marketed as an oxyalkylated cyclic tetramer was, in fact, a mixture of oxyalkylated cyclic octamer and cyclic hexamer. It is now realized that the *p*-substituted phenols generally do *not* form calix[4]arenes as the sole product but give mixtures that in most cases are exceedingly difficult to separate and that may or may not contain any cyclic tetramer. The idea at the outset of the Washington University project that this would be a neat, clean, and general way for constructing molecular baskets gave way at this point to a more pessimistic view of the reaction.

Out of adversity often comes advance, however, and the explorations of the *p-tert*-butylphenol–formaldehyde reaction proved this platitude. Further investigation of the one-step process by the St Louis group, involving changes in solvents, bases, reactant ratios, and other reaction variables, resulted in recipes that now permit the cyclic tetramer, cyclic hexamer, and cyclic octamer from *p-tert*-butylphenol to be easily prepared in good yield. These three compounds are among the most accessible synthetic macrocyclic

[53] C. D. Gutsche, A. E. Gutsche, and A. I. Karaulov, *J. Inclusion Phenom.*, **1985**, *3*, 447.

baskets, and they provide the starting point for a significant fraction of the phenol-derived calixarene chemistry that is being carried out in the world today. Phenol-derived calixarenes, born in Zinke's laboratory in 1941 out of the resinous tars that had been introduced to the world by Baeyer and Baekeland but largely unattended for the next thirty years, came of age as glistening crystalline solids in the 1970's mainly through the effort of Kämmerer and his group in Mainz, the group in Parma led by Ungaro, Andreetti, and Pochini, whose contributions have yet to be discussed, and Gutsche and his group in St Louis. Concomitantly, the resorcinol-derived calixarene octols were also coming of age through the pioneering efforts of Erdtman and Högberg in

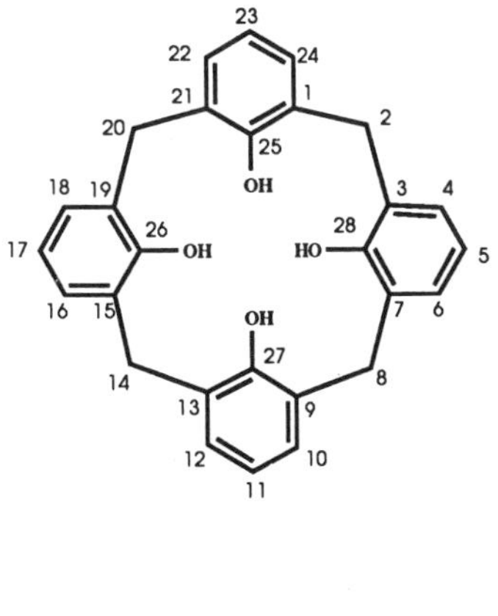

25,26,27,28-tetrahydroxycalix[4]arene

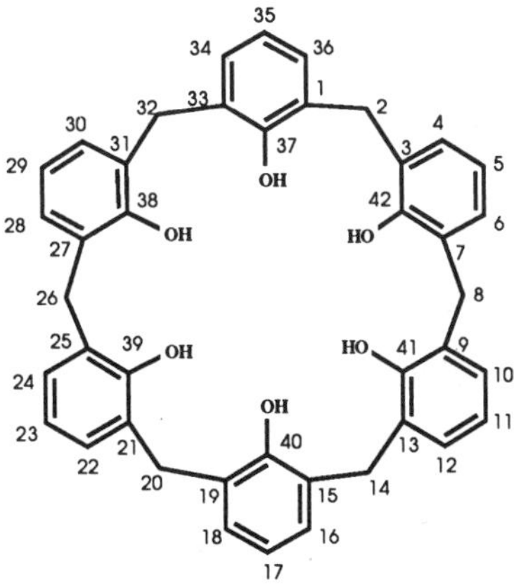

36,37,38,39,40,41,42-hexahydroxycalix[6]arene

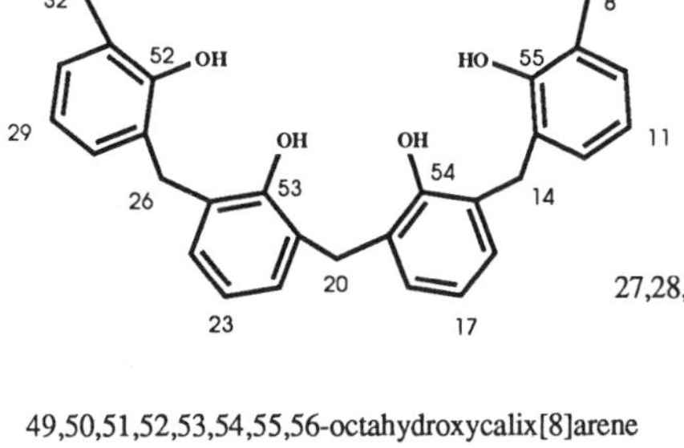

49,50,51,52,53,54,55,56-octahydroxycalix[8]arene

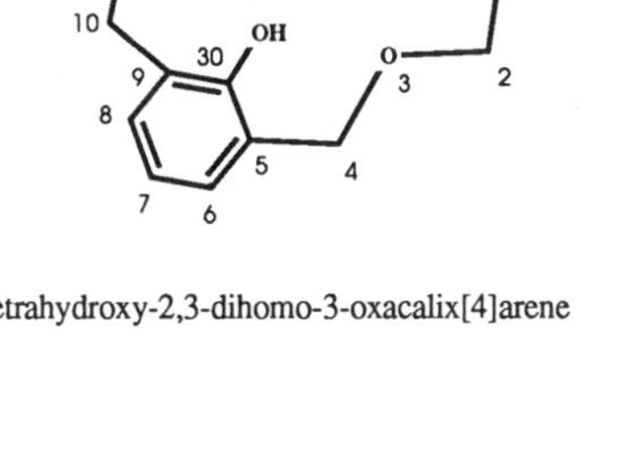

27,28,29,30-tetrahydroxy-2,3-dihomo-3-oxacalix[4]arene

Figure 1.5 *Structures and numbering of calix[4]arenes, calix[6]arenes, calix[8]arenes, and dihomooxacalix[4]arenes*

Stockholm. In the following chapters of this book we shall provide a detailed picture of the subsequent developments in calixarenes, a field of chemistry that is reaching a crescendo in the late 1980's.

1.8 Nomenclature of the Calixarenes

The name 'calixarene' was originally conceived to connote the shape of the phenol-derived cyclic tetramer in the conformation in which all four aryl groups are oriented in the same direction. To accommodate the name to the subsequently discovered oligomers containing more than four aryl groups a bracketed number is inserted between calix and arene. The product from the Petrolite Procedure, for example, is a calix[8]arene. Then, to indicate from which phenol the calixarene is derived, the *p*-substituent is designated by name. The cyclic tetramer from *p-tert*-butylphenol, for example, is called *p-tert*-butylcalix[4]arene. Resorcinol-derived calixarenes require a slight modification of nomenclature and are named 'calix[*n*]resorcinarenes'. The substituent at the methylene carbons (introduced by the aldehyde) is indicated by a prefix 'C-substituent'. The resorcinol-derived compound from *p*-bromobenzaldehyde, for example, is named as C-*p*-bromophenyl-calix[4]resorcinarene. For more systematic application of the calixarene nomenclature (*e.g.* in journal publications) the basic name 'calix[*n*]arene' is retained, and the identities of all substituents are indicated and their positions specified by the numbers shown in Figure 1.5.[54]

---

[54] Compound (**1**) (from *p-tert*-butylphenol and formaldehyde) is named 5,11,17,23-tetra-*tert*-butyl-25,26,27,28-tetrahydroxycalix[4]arene. Compound (**3**) (from resorcinol and acetaldehyde) is named 2,8,14,20-tetramethyl-4,6,10,12,16,18,22,24-octahydroxycalix[4]arene.

CHAPTER 2

Making the Baskets: Syntheses of Calixarenes

'We figure to ourselves
The thing we like, and then we build it up
As chance will have it, on the rock or sand'

Sir Henry Taylor, *Philip van Artevelde*, 1834

Creation myths in various cultures put basketmaking among the first of the arts given to humankind, a result of *homo sapiens'* innate desire to surround and transport objects. The present book deals with a modern facet of this ancient art and focuses on calixarenes as providing particular examples of how the chemist can build molecular baskets. As will be seen in later chapters, calixarenes are baskets that indeed can serve as vehicles of transport for chemical baggage such as metal ions and molecules. Before discussing these interesting properties, however, it is necessary first to learn how to make the baskets, and it is with this topic that the present chapter is concerned.

The previous chapter silhouetted the tortuous byways of phenol–formaldehyde chemistry that led in the 1970's to an unraveling of some of the apparent contradictions in the literature. One of the most useful results was the emergence of recipes by which the resinous tar can be induced to yield glistening crystals of various-sized macrocyclic compounds.

2.1 Procedures for the One-step, Base-induced Synthesis of Phenol-derived Calixarenes

2.1.1 Synthesis of *p-tert*-Butylcalixarenes

For many years the preparation of *p-tert*-butylcalix[4]arene remained a capricious event. Workable yields of the desired product were obtained in some instances, but poor yields or even no yields in other instances in reactions that were presumably carried out under identical conditions. The reasons for this variability still remain puzzling, but a careful investigation of the influence of varying amounts of base catalyst at various temperatures[1] has led to a set of instructions that are sufficiently reliable to insure easy

---

[1] C. D. Gutsche, M. Iqbal, and D. Stewart, *J. Org. Chem.*, **1986**, *51*, 742.

reproducibility.[2] One procedure, originally formulated by Zinke[3] and modified by Cornforth[4] and Gutsche,[1,2] is described as follows:

Modified   Zinke–Cornforth   Procedure   (Preparation   of   *p-tert*-butylcalix[4]arene): A mixture of *p-tert*-butylphenol, 37% formaldehyde, and an amount of NaOH corresponding to 0.045 equivalents with respect to the phenol is heated for 2 h at 110—120 °C to produce a thick viscous mass called the 'precursor'. The 'precursor' is then heated in refluxing diphenyl ether for 2 h, the reaction mixture is cooled, the crude product is separated by filtration and then recrystallized from toluene to give *ca.* 50% of glistening, white rhombs with mp 342—344 °C.

p-tert-Butylcalix[4]arene

In the procedure described by Zinke a 'neutralization' of what is referred to above as the 'precursor' removes some, but not all, of the base which is initially added to effect the condensation. The variability of results from one run to another using the original Zinke procedure, therefore, is probably due to the differing amount of base removed at this point. A neutron activation analysis[5] of a 'neutralized precursor' revealed that even after extensive trituration some sodium ion was retained, presumably as the phenoxide. Thus, exhaustive washing removes too much base and leads to little or no reaction, whereas brief washing leaves too much base and changes the course of the reaction in a manner described below. This ambiguity in the amount of base that is present during the 'thermal phase' can be eliminated by adding a known, calculated quantity at the beginning of the reaction and retaining it throughout the entire process, as described above.

There is an optimum amount of NaOH for the production of the calix[4]arene as indicated by the data in Figure 2.1 which show that the yield of product reaches a maximum at *ca.* 0.03—0.04 equivalents of base and falls off on either side of this range. With less base the yield of cyclic tetramer falls,

[2] C. D. Gutsche and M. Iqbal, *Org. Syn.*, forthcoming volume.
[3] A. Zinke and E. Ziegler, *Ber.*, **1941**, *B74*, 1729; *ibid.*, **1944**, *77*, 2644.
[4] J. W. Cornforth, P. D'Arcy Hart, G. A. Nicholls, R. J. W. Rees, and J. A. Stock, *Br. J. Pharmacol.*, **1955**, *10*, 73.
[5] C. D. Gutsche and D. Stewart, unpublished work.

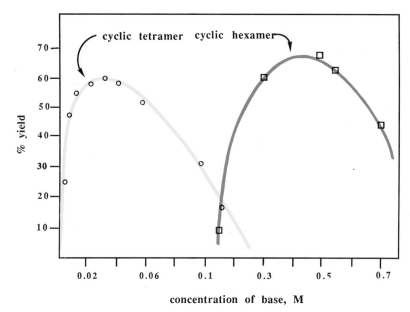

concentration of base, M

Figure 2.1   *Effect of base concentration on the formation of* p-tert-*butylcalix[4]arene*

ultimately to zero. With more base the yield of cyclic tetramer again falls, but to an increasing extent the product becomes cyclic hexamer. With 0.30 or more equivalents of base, cyclic hexamer is essentially the sole product. It must be admitted, however, that the realization of this fact came initially not from so rational a study as the one represented by these data but by chance. Early in the reinvestigation of the Petrolite Procedure[6] an amount of base ten times the intended quantity was added to the reaction mixture as the result of a weighing error, leading to a product that was ultimately identified as *p-tert*-butylcalix[6]arene. It is this accident that established the basis for a good preparation of the cyclic hexamer.

The amount of base that is used in the cyclooligomerization process affects the outcome in a profound fashion, while the cation that accompanies the basic anion has a smaller, though sometimes significant, effect[5,7] as follows: (*a*) LiOH is an inferior catalyst for the cyclooligomerization in general; (*b*) NaOH tends to give higher yields of cyclic octamer; (*c*) KOH, RbOH, and CsOH tend to give somewhat higher yields of cyclic hexamer, all other things being equal. This information provides the basis for a useful procedure for synthesizing the cyclic hexamer which is described as follows:

Modified Petrolite Procedure (Preparation of *p-tert*-Butylcalix[6]arene): A mixture of *p-tert*-butylphenol, 37% formaldehyde, and an amount of KOH

[6] C. D. Gutsche, B. Dhawan, K. H. No, and R. Muthukrishnan, *J. Am. Chem. Soc.*, **1981**, *103*, 3782. *idem, ibid.*, **1984**, *106*, 1891.
[7] B. Dhawan, S.-I. Chen, and C. D. Gutsche, *Makromol. Chem.*, **1987**, *188*, 921.

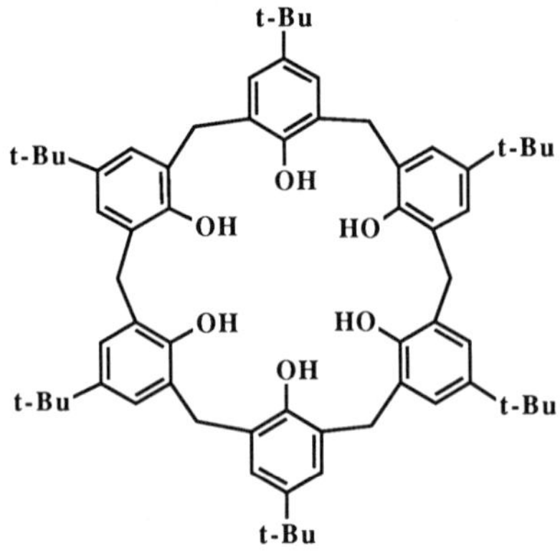

p-tert-Butylcalix[6]arene

corresponding to 0.34 equivalents with respect to the phenol is heated for 2 h in the manner described above to yield a light yellow, taffy-like 'precursor' which is then added to xylene and refluxed for 3 h. Filtration of the cooled reaction mixture yields a crude product which is neutralized and then recrystallized from chloroform—methanol to give a 80–85% yield of a white powder with mp 380—381 °C.

The best procedure for obtaining the cyclic octamer closely resembles that outlined in the Petrolite patents and is described as follows:

Standard Petrolite Procedure (Preparation of *p-tert*-butylcalix[8]arene): A slurry in xylene containing *p-tert*-butylphenol, paraformaldehyde, and an amount of NaOH corresponding to 0.03 equivalents with respect to the

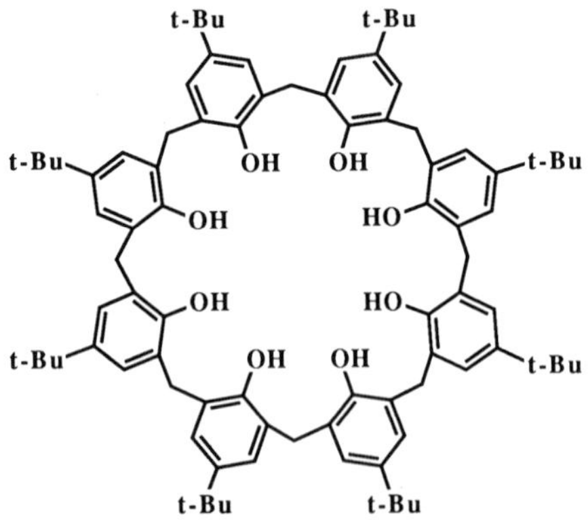

p-tert-Butylcalix[8]arene

phenol is refluxed for 4 h. The cooled reaction mixture is filtered, and the crude product is recrystallized from chloroform to give 60—65% of glistening crystals which quickly change to a white powder with mp 411—412 °C.

All three of the procedures described above can be easily carried out in the laboratory on a fairly large scale, making the *p-tert*-butylcalix[4], [6], and

[8]arenes readily available in reasonably large quantities. Far less readily accessible are the companion compounds *p-tert*-butylcalix[5]arene and *p-tert*-butylcalix[7]arene which are isolable only in much lower yields. Employing potassium *tert*-butoxide as the base, tetralin as the solvent, and a heating sequence consisting of 6 h at 55 °C and 6 h at 150 °C, Ninagawa and Matsuda[8] obtained a mixture from which they isolated, *inter alia*, 5% of the cyclic pentamer. Employing the Petrolite procedure but with dioxane as the solvent with a 30 h heating period, Nakamoto and Ishida[9] obtained a mixture from which they isolated 6% of the cyclic heptamer. The yield of the latter has been increased to *ca.* 20% in St Louis laboratories[10] by using an amount of base corresponding to one full equivalent per phenolic hydroxyl group. It is unfortunate that the calix[5]arene is so difficult to obtain, because it is particularly attractive for enzyme mimic building. *p-tert*-Butylcalix[9]arene has recently been isolated[5] from a reaction mixture obtained by a modified Petrolite procedure, and there is mass spectral and HPLC evidence for a *p-tert*-butylcalix[10]arene and even larger cyclooligomers.

2.1.2 Synthesis of Other *p*-Substituted Calixarenes

While *p-tert*-butylphenol is the quintessential reactant in the base-induced calixarene-forming reaction, it is not the only phenol that behaves in this fashion. *p-tert*-Pentylphenol[11] and *p*-(1,1,3,3-tetramethylbutyl)phenol[4,12,13] (often referred to as *p-tert*-octylphenol) both react rather comparably to *p-tert*-butylphenol, although the products are somewhat less tractable and the yields somewhat lower. Vicens and coworkers[14] report the conversion of *p*-isopropylphenol to the cyclic octamer, but the St Louis group[7] finds this phenol to be less satisfactory than those containing quaternary carbons directly attached to the phenyl ring. *p*-Cresol gives a product that is high melting and insoluble in organic solvents, characteristic of a calixarene. Analysis by TLC indicates the presence of several compounds, and osmometry gives a molecular weight that is even higher than that expected for a cyclic octamer. The condensation of formaldehyde with *p*-phenylphenol, already alluded to as a potential source of a calixarene of particular interest because of its deep cavity, has been studied in some detail.[15] Each of the one-step procedures for making calixarenes yields mixtures from *p*-phenylphenol that contain little or none of the cyclic tetramer, an authentic sample of *p*-phenylcalix[4]arene being used to demonstrate its complete absence in any of these reaction mixtures.[15] Instead, the products are *p*-phenylcalix[6]arene,[15]

[8] A. Ninagawa and H. Matsuda, *Makromol. Chem. Rapid Commun.*, **1982**, *3*, 65.

[9] Y. Nakamoto and S. Ishida, *Makromol. Chem. Rapid Commun.*, **1982**, *3*, 705.

[10] C. D. Gutsche and I. Alam, unpublished observations.

[11] S. R. Izatt, R. T. Hawkins, J. J. Christensen, and R. M. Izatt, *J. Am. Chem. Soc.*, **1985**, *107*, 63.

[12] J. W. Cornforth, E. D. Morgan, K. T. Potts, R. J. W. Rees, *Tetrahedron*, **1973**, *29*, 1659.

[13] V. Bocchi, D. Foina, A. Pochini, and R. Ungaro, *Tetrahedron*, **1982**, *38*, 373.

[14] J. Vicens, T. Pilot, D. Gamet, R. Lamartine, and R. Perrin, *C. R. Acad. Sci. Paris*, **1986**, *302*, 15.

[15] C. D. Gutsche and P. F. Pagoria, *J. Org. Chem.*, **1985**, *50*, 5795.

p-phenylcalix[7]arene,[16] and *p*-phenylcalix[8]arene.[15] Since the phenols that seem most amenable to smooth calixarene production are those carrying a quaternary carbon directly attached to the *p*-position of the phenolic ring (*e.g. p-tert*-butyl-, *p-tert*-pentyl-, and *p-tert*-octyl-phenols) it was thought that *p*-adamantylphenol should behave in a comparable fashion. It does, indeed, react with formaldehyde, but the very high melting product that is formed is so insoluble and intractable that characterization has not yet been successful.[17]

Recently, a number of phenols carrying *n*-alkyl groups in the *p*-position have been converted into calixarenes. Nakamoto and coworkers[18] report that *p-n*-octylphenol, *p-n*-nonylphenol, and *p-n*-dodecylphenol yield the corresponding calixarenes as mixtures of the cyclic hexamer, heptamer, and octamer. Asfari and Vicens[19] report the preparation of calix[6]arenes and calix[8]arenes from phenols carrying *p*-alkyl groups with 8, 10, 14, 16, and 18 carbon atoms. The experimental details in these reports give little information about the yield and purity of the products, however, so it is difficult to assess their practicality. The scope of the base-induced, one-step preparation of phenol-derived calixarenes can thus be summarized as follows: (*a*) only *p*-isopropyl, *p-tert*-butyl, *p-tert*-pentyl, and *p*-(1,1,3,3-tetrabutyl)phenol are known with certainty to give cyclic tetramers; (*b*) these four phenols as well as a number of others, including *p*-phenylphenol and *p-n*-alkylphenols, produce cyclic hexamers, heptamers, and/or octamers, generally as mixtures; (*c*) *p-tert*-butylphenol is generally the compound of choice for the preparation of the cyclic tetramer, hexamer, or octamer.

2.2   Procedures for the One-step, Acid-catalyzed Synthesis of Calixarenes

2.2.1   Phenol-derived Calixarenes

Although pure calixarenes have not been isolated from acid-catalyzed reactions of phenols with formaldehyde or higher aldehydes, there is evidence that they are present in small amounts[5,20] in the crude reaction mixtures.

2.2.2   Resorcinol-derived Calixarenes

A product from the acid-catalyzed reaction of resorcinol and formaldehyde has been assigned a linear dimeric structure,[21] but its non-crystalline and

[16] Jpn. Kokai Tokkyo Koho, **1984**, JP 59,104,331 (*Chem. Abstr.*, 101, 191410v); *idem*, **1984**, JP 59,104,332 (*Chem. Abstr.*, 101, 191,409b); *idem*, **1984**, JP 59,104,333 (*Chem. Abstr.*, 101, 19141w).

[17] S.-I. Chen, PhD Thesis, Washington University, St Louis, **1984**.

[18] Y. Nakamoto, T. Kozu, S. Oya, and S. Ishida, *Netsu Kokasei Jushi*, **1985**, 6, 78 (*Chem. Abstr.*, 105, 6301g).

[19] Z. Asfari and J. Vicens, *Tetrahedron Lett.*, **1988**, 29, 2659.

[20] Dr. F. J. Ludwig, unpublished observations.

[21] N. Caro, *Ber.*, **1892**, 25, 939.

refractory character (decomposes at 250 °C without melting) suggests that it is actually a polymer. Resorcinol and formaldehyde are known to react under both acid-catalyzed or base-induced conditions to form polymers that have had long commercial use as adhesives. Less reactive aldehydes, however, do condense with resorcinol to give cyclic tetramers, as discussed in Chapter 1. Subsequent to the work of Niederl in 1940 considerable attention has been given to the resorcinol-derived calixarenes, and their structures have been conclusively proved by the methods discussed in the following chapter. Högberg in particular has studied them in detail[22-24] and has published a procedure that has been adapted to large-scale operation by Cram and coworkers[25] for the preparation of calix[4]resorcinarenes.

Neiderl–Högberg Procedure (Preparation of C-methylcalix[4]resorcinarene): A solution of resorcinol and acetaldehyde in aqueous ethanol is maintained at 80 °C for 16 h. The cooled reaction mixture is filtered to give *ca.* 70% of a light yellow product that is pure enough for subsequent use.

C-Methylcalix[4]resorcinarene

It is quite remarkable that a single compound is formed in such high yield in the resorcinol–aldehyde reaction, because conformational and configurational features make possible a rather large number of diastereoisomers. A particularly useful piece of information from Högberg's studies is the fact that the system is dynamic, equilibration among the diastereoisomers causing the product composition to be a function of time as illustrated in Figure 2.2.[24] Isomer-A is the major product after 30 min and isomer-B virtually the sole product after 10 h, a classic example of kinetic control *vs.* thermodynamic control. A third isomer has been isolated in very low yield from the reaction of resorcinol with heptaldehyde or dodecylaldehyde.[26]

The resorcinol–aldehyde reaction is amenable to structural variations in both of the reactants. Among the aldehydes that have been used are

[22] A. G. S Högberg, Ph.D. dissertation, Royal Institute of Technology, Stockholm, **1977**. Appreciation is expressed to Dr. Högberg for providing a copy of this excellent thesis.
[23] A. G. S. Högberg, *J. Org. Chem.*, **1980**, *45*, 4498.
[24] A. G. S. Högberg, *J. Am. Chem. Soc.*, **1980**, *102*, 6046.
[25] D. J. Cram, S. Karbach, H. E. Kim, C. B. Knobler, E. F. Maverick, J. L. Ericson, and R. C. Helgeson, *J. Am. Chem. Soc.*, **1988**, *110*, 2229.
[26] L. Abis, E. D. Dalcanale, A. Du vosel, and S. Spera, *J. Org. Chem.*, **1988**, *53*, 5475.

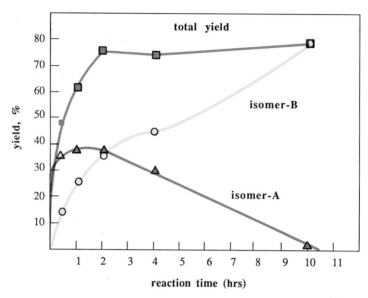

Figure 2.2 *Acid-catalyzed reaction of resorcinol and benzaldehyde[24]*

acetaldehyde,[25,27] propionaldehyde,[27] isobutyraldehyde,[28] isovalderaldehyde,[27] heptaldehyde,[26] dodecylaldehyde,[26] benzaldehyde,[24,28] *p*-bromobenzaldehyde,[28] and ferrocenecarboxaldehyde.[29] In addition to resorcinol, 2-methylresorcinol and 2-bromoresorcinol react with acetaldehyde to afford the corresponding C-methylcalix[4]resorcinarenes.[25]

Acid-catalyzed reactions of phenols and aldehydes produce mixtures of linear oligomers, some of considerable length,[30] the tendency to cyclize obviously being far lower than in the base-induced process. Ketones, also, fail to form calixarenes from phenols under either acidic or basic conditions. It is interesting to note, though, that non-hydroxyl-containing calixarenes have been made by the acid-catalyzed condensation of a variety of aromatic compounds. For example, mesitylene and 1,2,3,5-tetramethylbenzene both react with formaldehyde in the presence of acetic acid to yield cyclic tetramers **1** ($R^1 = H$, $R^2 = Me$) and **1** ($R^1 = R^2 = Me$).[31] A Friedel–Crafts reaction with chloromethylmesitylene[32] leads to the same result. This latter procedure has

[27] J. B. Niederl and H. J. Vogel, *J. Am. Chem. Soc.*, **1940**, *62*, 2512.
[28] H. Erdtman, S. Högberg, S. Abrahamsson, and B. Nilsson, *Tetrahedron.Lett.*, **1968**, 1679.
[29] P. D. Beer and E. L. Tite, *Tetrahedron Lett.*, **1988**, *29*, 2349.
[30] W. J. Burke, W. E. Craven, A. Rosenthal, S. H. Ruetman, C. W. Stephens, and C. Weatherbee, *J. Polymer Sci.*, **1956**, *20*, 75.
[31] J. L. Ballard, W. B. Kay, and E. L. Kropa, *J. Paint Technol.*, **1966**, *38*, 251.
[32] F. Bottino, G. Montaudo, and P. Maravigna, *Ann. Chim. (Rome)*, **1967**, *57*, 972.

been extended to the use of a methoxylated starting compound which produces an 80% yield of calixarene 1 ($R^1 = H$, $R^2 = OMe$) carrying four extraannular methoxyl groups,[33] as illustrated in Figure 2.3. In a closely related process, an acid-catalyzed reaction of 2-methoxyazulene with formaldehyde is reported[34] to afford a 5% yield of the tetramethyl ether of azulocalix[4]arene (see Figure 2.11).

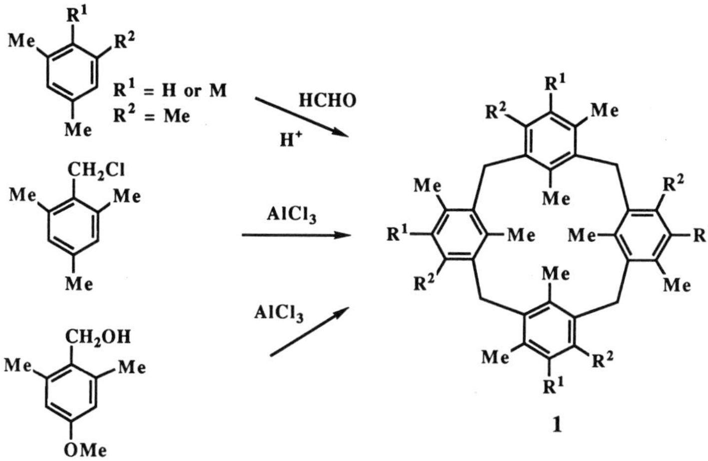

Figure 2.3 *One-step, acid-catalyzed synthesis of non-hydroxylic calix[4]arenes*[30-32]

2.3 Procedures for the One-step Formation of Calixarenes under Neutral Conditions

Although almost all of the condensations of phenols and aldehydes described in the literature are acid-catalyzed or base-induced, there are a few examples of reactions under neutral conditions. As part of an extensive program in phenol–formaldehyde chemistry, Casiraghi and Casnati and their co-workers[35] have shown that linear oligomers can be obtained from a variety of phenols reacting with formaldehyde under neutral conditions. Chasar[36] has provided a calixarene example by isolating compound 3 in 20% yield from 2,2'-di-*tert*-butyl-4,4'-diisopropylidene-bis(phenol) (2) and paraformaldehyde heated in xylene at 175 °C in an autoclave for 12 h, as illustrated in Figure 2.4. *p-tert*-Butylphenol and formaldehyde show only traces of calixarene when heated in refluxing diphenyl ether.[10]

[33] T.-T. Wu and J. R. Speas, *J. Org. Chem.*, **1987**, *52*, 2330.
[34] T. Asao, S. Ito, and N. Morita, *Tetrahedron Lett.*, **1988**, *29*, 2839.
[35] G. Casiraghi, G. Casnati, M. Cornia, G. Sartori, and F. Bigi, *Makromol. Chem.*, **1981**, *182*, 2973.
[36] D. W. Chasar, *J. Org. Chem.*, **1985**, *50*, 545.

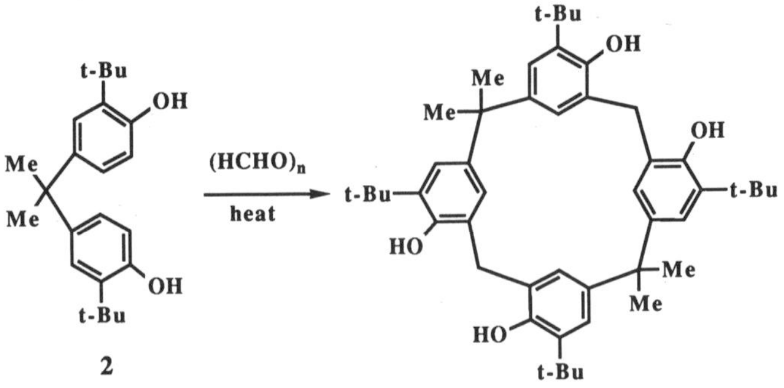

Figure 2.4 *Thermally-induced one-step synthesis of a calix[4]arene[36]*

2.4 Multi-step Synthesis of Calixarenes

2.4.1 Non-convergent Stepwise Syntheses

The 10-step procedure employed by Hayes and Hunter[37] for the synthesis of *p*-methylcalix[4]arene was discussed in Chapter 1 (see Figure 1.1). It represents a completely linear, non-convergent approach which, though long and tedious, in principle is amenable to considerable variation through the use of different phenols at each of the individual arylation steps. It is an approach that has been exploited and improved in impressive fashion by Kämmerer and his group at the University of Mainz in Germany. Hermann Kämmerer was born in 1911 in Milan and received his PhD degree at the University of Fribourg under the direction of Hermann Staudinger (Nobel Laureate in 1953). His doctoral thesis had already kindled an interest in polymer chemistry, and in 1947 when he joined the staff at the University of Mainz he started work on phenol–formaldehyde resins with the particular aim of discerning their structure. To this end he undertook the synthesis of a series of well-defined oligomers. At the outset all of these were linear, although in a paper published in 1962 reference is made to the cyclic tetramers of Zinke. Eventually, his synthetic efforts turned to the cyclics, which led to a series of elegant and detailed papers.[38-45] In the first paper of the series the Hayes and Hunter synthesis was repeated essentially without change, leading to *p*-methylcalix[4]arene(**5**) as shown in Figure 2.5. In the

[37] B. T. Hayes and R. F. Hunter, *Chem. Ind.*, **1954**, 193; *idem, J. Applied Chem.*, **1958**, *8*, 743.
[38] H. Kämmerer, G. Happel, and F. Caesar, *Makromol. Chem.*, **1972**, *162*, 179.
[39] G. Happel, B. Mathiasch, and H. Kämmerer, *Makromol. Chem.*, **1975**, *176*, 3317.
[40] H. Kämmerer and G. Happel, *Makromol. Chem.*, **1978**, *179*, 1199.
[41] H. Kämmerer, G. Happel, V. Böhmer, and D. Rathay, *Monatsh. Chem.*, **1978**, *109*, 767.
[42] H. Kämmerer and G. Happel, *Makromol. Chem.*, **1980**, *181*, 2049.
[43] H. Kämmerer and G. Happel, *Monatsh. Chem.*, **1981**, *112*, 759.
[44] H. Kämmerer, G. Happel, and B. Mathiasch, *Makromol. Chem.*, **1981**, *182*, 1685.
[45] H. Kämmerer and G. Happel in '*Weyerhaeuser Science Symposium on Phenolic Resins, 2*', Tacoma, Washington, 1979, Weyerhaeuser Publishing Co., Tacoma, **1981**, p. 143.

Hermann Kämmerer

7

p-hexamethyl-*p*-*tert*-butylcalix[7]arene

5

p-methylcalix[4]arene

Figure 2.5 *Stepwise synthesis of a calix[4]arene and a calix[7]arene*[42]

light of subsequent developments in calixarene chemistry, perhaps the most significant contribution in this paper was the observation of the temperature-, dependent character of the ^{1}H NMR resonances arising from the bridging CH$_2$ groups, a feature that has been of great importance in studying the conformational behavior of these compounds (see Chapter 4).

The Hayes and Hunter method, of course, is not limited to the synthesis of cyclic tetramer, for mono-hydroxymethyl linear oligomers of other lengths have the potential for cyclization to the corresponding cyclooligomer. That this is true has been nicely demonstrated by Kämmerer and colleagues in syntheses of cyclooligomers containing five, six, and even seven aromatic residues. In a *tour de force* reaction sequence whose repetitive nature is reminiscent of polypeptide and polynucleotide syntheses, Kämmerer and Happel[42] prepared the calix[7]arene (7) carrying six *p*-methyl groups and one *p-tert*-butyl group. The synthesis, adumbrated in Figure 2.5 starting with *p*-cresol, is a 16-step process that includes six arylations with yields ranging from 56—76% (averaging *ca.* 72%), seven hydroxymethylations with yields ranging from 46—82% (averaging *ca.* 63%), bromination and debromination, and a final cyclization which proceeds in 62% yield (higher than the 46% yield reported in the synthesis of the cyclic tetramer). Employing these same procedures and using *p*-cresol and *p-tert*-butylphenol as the arylating moieties, they prepared the thirteen *p*-alkyl-calixarenes (8—10) shown in Table 2.1. It is interesting to note, though, that the Zinke products were so confidently thought to be cyclic tetramers that no comparison was ever made between Kämmerer's authentic compounds and those produced in the Zinke reaction. This was not accomplished until 1981 when Gutsche and coworkers[6] prepared *p-tert*-butylcalix[4]arene by the Hayes and Hunter route and showed that it was identical with one of the materials that can be obtained by the base-induced condensation of *p-tert*-butylphenol and formaldehyde.

Still another example of a Hayes and Hunter synthesis is that of *p*-phenylcalix[4]arene, prepared by Gutsche and No[46] for investigating the composition of the product from the one-step reaction of *p*-phenylphenol and formaldehyde (see p. 45—6). This synthesis proved to be less facile than that of the *p*-alkyl-substituted calixarenes, and the overall yield was only *ca.* 0.5% as compared with *ca.* 11% for *p-tert*-butylcalix[4]arene. Furthermore, it was complicated by the formation of three compounds in the cyclization reaction of 11, as shown in Figure 2.6. One of these is the desired product (12), and the other two are thought to be the isomeric compounds 13 and 14 formed as the result of electrophilic substitution at the other reactive sites of the *p*-phenylphenol moiety at the end of the linear tetramer.

2.4.2   Convergent Stepwise Syntheses

The Hayes and Hunter stepwise synthesis is long and rather tedious. Although its individual steps usually proceed reasonably well, the overall

[46] K .H. No and C. D. Gutsche, *J. Org. Chem.*, **1982**, *47*, 2713.

Table 2.1 *Calixarenes synthesized by Kämmerer and coworkers* via *the Hayes and Hunter stepwise method*

Substituent

Cmpd	R^1	R^2	R^3	R^4	R^5	R^6	Ref
8a	H	H	Me	Me			42
b	Me	Me	Me	Me			39
c	Me	Me	Me	t-Bu			41
d	Me	Me	t-Bu	t-Bu			41,42
e	Me	t-Bu	Me	t-Bu			41
f	Me	t-Bu	t-Bu	t-Bu			39,41
g	t-Bu	t-Bu	t-Bu	t-Bu			40—42

Cmpd	R^1	R^2	R^3	R^4	R^5	R^6	Ref
9a	Me	Me	Me	Me	Me		40,43
b	Me	Me	Me	Me	t-Bu		45
c	Me	Me	t-Bu	t-Bu	t-Bu		45

Cmpd	R^1	R^2	R^3	R^4	R^5	R^6	Ref
10a	Me	Me	Me	Me	Me	Me	43
b	Me	Me	Me	Me	Me	t-Bu	43
c	t-Bu	Me	Me	Me	Me	Me	43

yields seldom exceed 10% and generally are lower. Recognizing these deficiencies, several research groups have devised convergent pathways that reduce the total number of steps and improve the synthetic utility of the stepwise method. Böhmer's group at the University of Mainz has been the most successful in developing and exploiting this approach. Volker Böhmer was born in Rauscha, Germany in 1941 and received his PhD training at the University of Mainz under the direction of Hermann Kämmerer. Already conversant with phenol–formaldehyde chemistry and the stepwise syntheses of linear as well as cyclic oligomers as a result for his doctoral work, Böhmer's entry into calixarene chemistry was more or less predestined. Eventually becoming a member of the faculty at the University of Mainz, he has inherited the mantle of his mentor and is carrying on a tradition that pays homage to detailed physical organic chemical investigation. To extend his

Figure 2.6   *Stepwise synthesis of p-phenylcalix[4]arene by the Hayes and Hunter method[46]*

Volker Böhmer

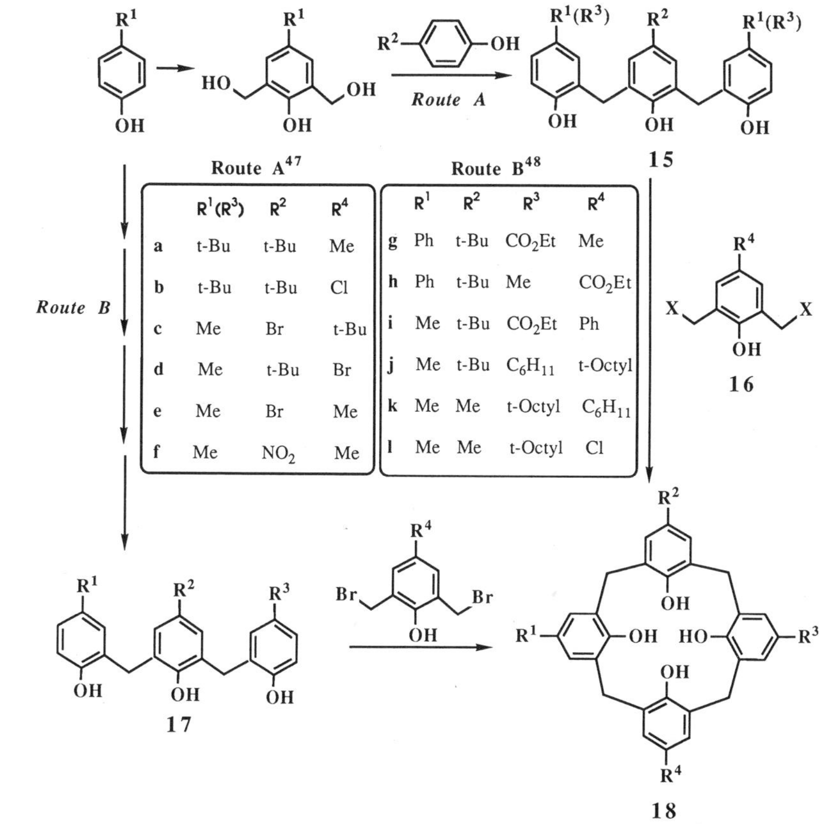

	$R^1(R^3)$	R^2	R^4		R^1	R^2	R^3	R^4
a	t-Bu	t-Bu	Me	g	Ph	t-Bu	CO_2Et	Me
b	t-Bu	t-Bu	Cl	h	Ph	t-Bu	Me	CO_2Et
c	Me	Br	t-Bu	i	Me	t-Bu	CO_2Et	Ph
d	Me	t-Bu	Br	j	Me	t-Bu	C_6H_{11}	t-Octyl
e	Me	Br	Me	k	Me	Me	t-Octyl	C_6H_{11}
f	Me	NO_2	Me	l	Me	Me	t-Octyl	Cl

Figure 2.7 *'3 + 1' Convergent stepwise synthesis of calix[4]arenes*[47, 48]

studies of hydrogen bond effects in polyphenolic systems Böhmer required cyclic oligomers with chemically more interesting functionality than those provided by the Kämmerer syntheses. It was this quest that led Böhmer, Chhim, and Kämmerer[47] to explore what can be called a '3 + 1' approach in which a linear trimer (**15** or **17**) is condensed with a 2,6-bis-halomethyl-phenol (**16**), as illustrated in Figure 2.7. The shorter route using this sequence (route A) involves the symmetrical linear trimer **15**, prepared by treating a bishydroxymethylphenol with an excess of a phenol. If an unsymmetrical linear trimer **17** is desired (route B),[48] the longer stepwise approach must be used, involving protection and deprotection. Although much shorter than the completely linear method, this convergent procedure gives rather low yields in the cyclization step, ranging from 10—20% in the best cases down to 2—7% in the more interesting cases in which hetero atoms and functional groups are included as *p*-substituents on the aryl moieties. The use of a large

[47] V. Böhmer, P. Chhim, and H. Kämmerer, *Makromol. Chem.*, **1979**, *180*, 2503.
[48] V. Böhmer, F. Marschollek, and L. Zetta, *J. Org. Chem.*, **1987**, *52*, 3200.

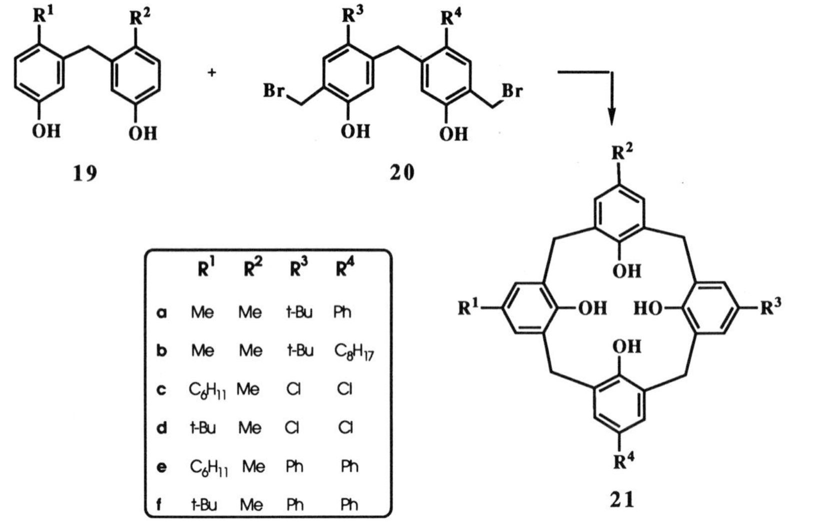

	R1	R2	R3	R4
a	Me	Me	t-Bu	Ph
b	Me	Me	t-Bu	C_8H_{17}
c	C_6H_{11}	Me	Cl	Cl
d	t-Bu	Me	Cl	Cl
e	C_6H_{11}	Me	Ph	Ph
f	t-Bu	Me	Ph	Ph

Figure 2.8 *'2 + 2' Convergent stepwise synthesis of calix[4]arenes*[49]

excess of $TiCl_4$ in the reaction mixture[48] somewhat improves the yields and has the advantage of obviating the necessity for high dilution, allowing the cyclizations to proceed under ordinary conditions.

A closely related scheme, shown in Figure 2.8, is the '2 + 2' approach which Böhmer[49] has also studied in detail. As in the previous case, a significant improvement in the synthesis results from the use of $TiCl_4$ in the cyclization step. The yields of **21** range from 9.5—21%, with most of them being quite close to 20%. The choice between the '3 + 1' and '2 + 2' methods appears to be primarily dependent on the relative ease of construction of the component parts, *viz.* **15** or **17** in the first case and **19** in the second. In an interesting extension of this general approach, Böhmer and coworkers[50–53] have prepared bridged calixarenes by what can be classed as a '2 + 1 + 1' approach from bisphenols (**22**) and bis-dibromomethylphenols (**23**), as illustrated in Figure 2.9. Viewing the products (**24**) as baskets, Böhmer named then 'arrichoarenes' (Greek *arriches*, 'basket with a handle') and equipped them with handles ranging in length from $n = 5$ to $n = 16$ in yields ranging from 2—20% and peaking at $n = 8$.

Using a '2 + 2' procedure similar to that described by Böhmer, a group in Iran has investigated the synthesis of a variety of *p*-halocalix[4]arenes in con-

[49] V. Böhmer, L. Merkel, and U. Kunz, *J. Chem. Soc., Chem. Commun.*, **1987**, 896.
[50] V. Böhmer, H. Goldmann, and W. Vogt, *J. Chem. Soc., Chem. Commun.*, **1985**, 667.
[51] E. Paulus, V. Böhmer, H. Goldman, and W. Vogt, *J. Chem. Soc., Perkin Trans. 2*, **1987**, 1609.
[52] V. Böhmer, H. Goldmann, R. Kaptein, and L. Zetta, *J. Chem. Soc., Chem. Commun.*, **1987**, 1358.
[53] H. Goldman, W. Vogt, E. Paulus, and V. Böhmer, *J. Am. Chem. Soc.*, **1988**, *110*, 6811.

Figure 2.9 *'2 + 1 + 1' Convergent stepwise synthesis of bridged calix[4]arenes[50−53]*

nection with a program dealing with analogs of phloroglucides.[54−56] They reported that the condensation of the dimer **25** (X = Cl) with the bis-hydroxy-methyl dimer **28** (X = Cl) affords a 65% yield of *p*-chlorocalix[4]arene (**29**, X = Cl). The properties that are described for this product are at such wide variance with those of an authentic sample of *p*-bromocalix[4]arene,[15] however, that there is some question whether the reaction has taken the course indicated in Figure 2.10. Also questionable is the contention that the

CPK Model of an arrichoarene

[54] A. A. Moshfegh, R. Badri, M. Hojjatie, M. Kaviani, B. Naderi, A. H. Nazmi, M. Ramezanian, B. Roozpeikar, and G. H. Hakimelahi, *Helv. Chim. Acta*, **1982**, *65*, 1221.

[55] A. A. Moshfegh, B. Mazandarani, A. Nahid, and G. H. Hakimelahi, *Helv. Chim. Acta*, **1982**, *65*, 1229.

[56] A. A. Moshfegh, E. Beladi, L. Radnia, A. S. Hosseini, S. Tofigh, and G. H. Hakimelahi, *Helv. Chim. Acta*, **1982**, *65*, 1264.

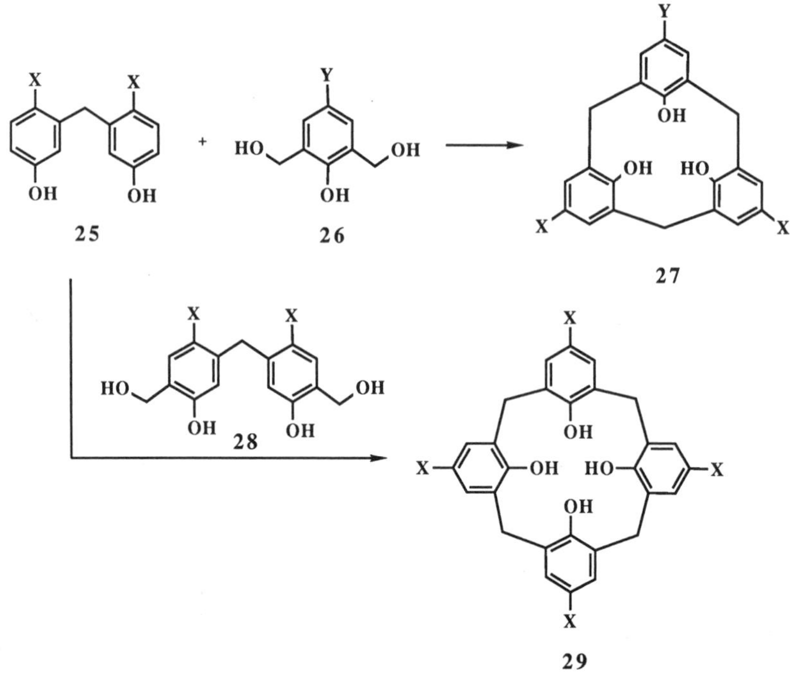

Figure 2.10   *'2 + 2' Convergent stepwise synthesis of calixarenes*[54–56]

reaction of **25** and **26** yields the calix[3]arene **27**. Attempts to duplicate this outcome have failed,[57] and the inability to construct **27** with CPK models suggests that it is probably too strained to be easily formed in a reaction of this type. It seems probable that some of the cyclizations reported by these workers have led to oxacalixarenes (see Section 2.7) rather than calixarenes.

Closely related to the synthesis shown in Figure 2.10 is the recently reported[34] preparation of an azulocalixarene by the acid-catalyzed condensation of **31** and **33**, obtained by the conversion of 2-methoxyazulene (**30**) to the linear dimer **31** followed by transformation to **33** either by formylation and reduction or carbethoxylation and reduction, as shown in Figure 2.11. Stirring a dilute CH_2Cl_2 solution containing equal amounts of **31** and **33** for 1 h in the presence of acetic acid yields 20% of the cyclic tetramer **34**, isolated as blue needles melting at 166–167 °C. Surprisingly, the calixarene **34** was also obtained in 10% yield when a 1:1 mixture of **30** and **33** was treated with acetic acid in a similar fashion and in 5% yield when **30** was treated with paraformaldehyde in the presence of acetic acid.

Still another approach to a convergent synthesis has been devised by No and Gutsche[46] which retains some of the functional group flexibility of the Hayes and Hunter method and affords reasonably good yields of calixarenes. It is a 4-step sequence, as illustrated in Figure 2.12, in which a *p*-substituted

---

[57] C. D. Gutsche, unpublished observations; V. Böhmer, personal communication.

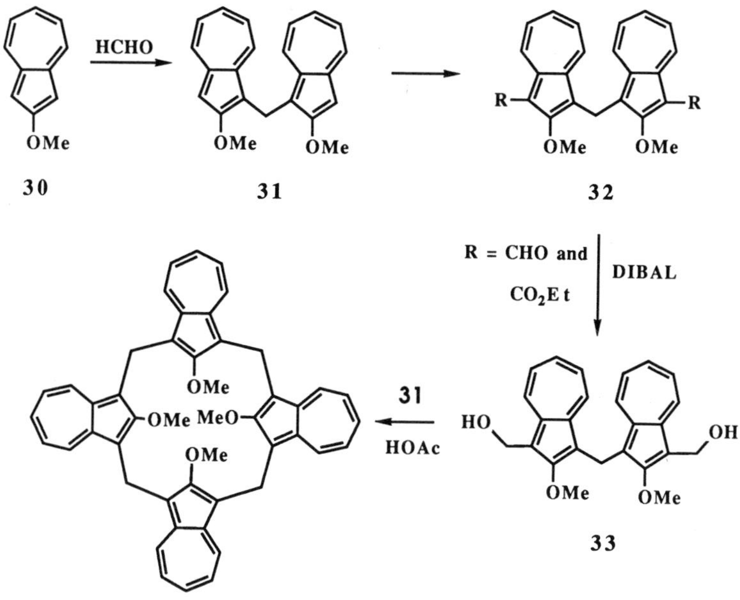

Figure 2.11   *'2 + 2' Convergent synthesis of an azulocalix[4]arene*[34]

phenol (*e.g. p*-phenylphenol) is treated with formaldehyde under carefully controlled conditions to produce the bis-hydroxymethyl dimer **35**. Acid-catalyzed arylation with two equivalents of a phenol yields **36** which is selectively hydroxymethylated to form **37** (Y = H). Acid-catalyzed cyclization using the standard Hayes and Hunter conditions then produces the calix[4]arene **38**. Although the overall yield is only *ca.* 10%, the starting materials are cheap and the work-up procedures are simple, the principal drawback in the method being the separation of **37** from non-hydroxymethylated compound (**36**) and *bis*-hydroxymethylated compound **37** (Y = CH₂OH). The convergent procedures that have been devised by Casiraghi and Casnati and their coworkers[58] for the preparation of linear oligomers also take on added interest in this context.

Uncatalyzed condensations leading to calixarenes have been achieved by the Parma group[59] who have shown that **15** (R¹ = R² = R³ = Me) reacts with **16** (R⁴ = Me, X = OH) (see Figure 2.7) at 150 °C in the presence of xylene to give a 20% yield of a mixture containing 30% of *p*-methylcalix[4]arene and 70% of *p*-methylcalix[5]arene. This surprising result in which the cyclic pentamer is the major product is rationalized in terms of partial deformylations (leading to monohydroxymethylated intermediates)

---

[58] E. Dradi, G. Casiraghi, G. Sartori, and G. Casnati, *J. Chem. Soc., Chem. Commun.*, **1978**, 1295.

[59] A. Arduini, A. Casnati, A. Pochini, and R. Ungaro, 5th International Symposium on Inclusion Phenomena and Molecular Recognition, Orange Beach, Ala, **1988**, p. H13.

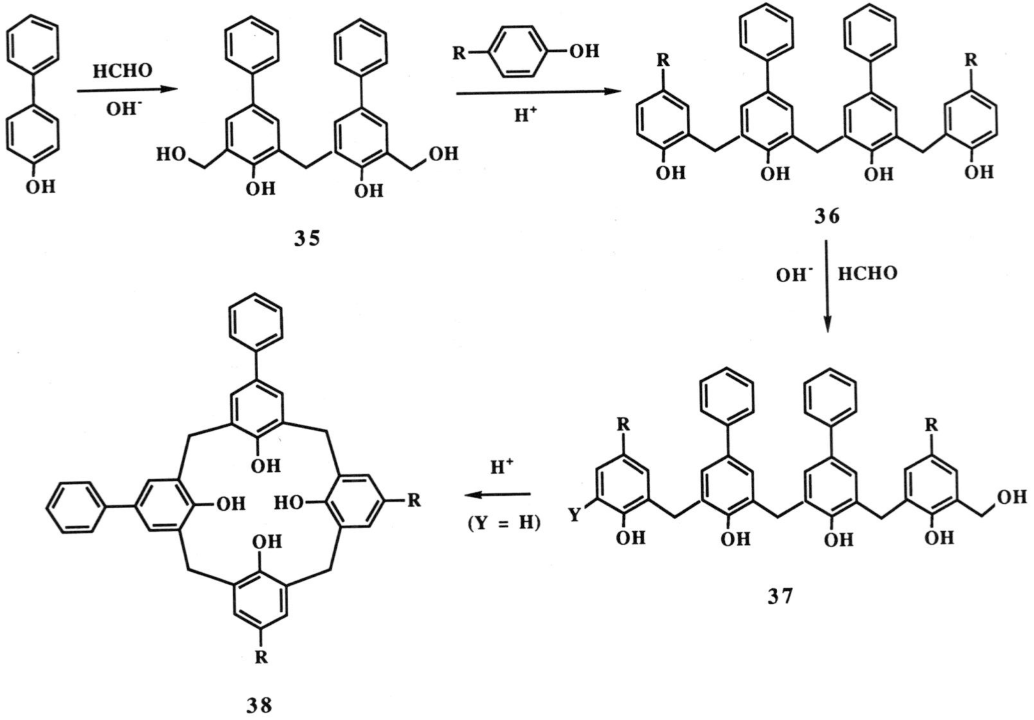

Figure 2.12 *A Hayes and Hunter-type '2 + 1 + 1' convergent stepwise synthesis of calix[4]arenes*[46]

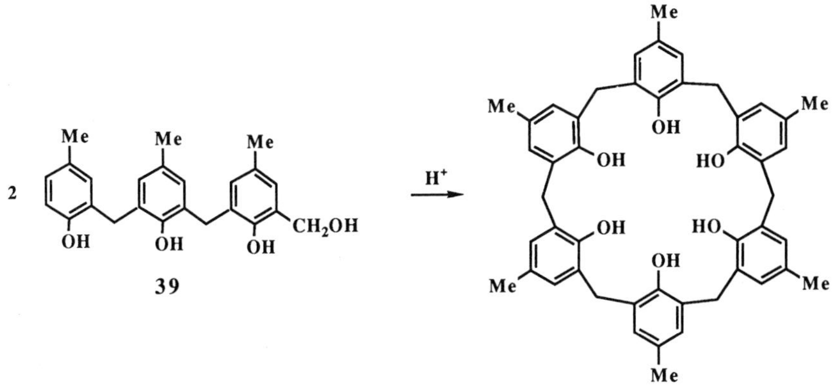

Figure 2.13 *'2 + 2' Convergent stepwise synthesis of calix[6]arenes*[43]

and kinetic competition between the formation of cyclic tetramer and cyclic pentamer.

Although almost all of the experiments with convergent stepwise methods have been directed to calix[4]arenes, there have been at least two attempts to extend the method to larger calixarenes. Kämmerer and Happel[43] report, though without experimental procedures, that the acid-catalyzed condensation of a pair of mono-hydroxymethylated linear trimers (**39**) produces the calix[6]arene **40** in low yield, as pictured in Figure 2.13. It will be interesting to learn whether the TiCl$_4$-catalyzed process that Böhmer has used so successfully in preparing the calix[4]arenes can be extended to the synthesis of larger calixarenes. Hunter and Turner[60] reported a 'rational synthesis of a cyclic octanuclear novolak' by acid-catalyzed condensation of equimolar portions of **41** and **42** to yield what they assumed to be **43** (R = CH$_2$) as shown in Figure 2.14. However, no experimental details for this synthesis were provided, and the product was characterized only by a molecular weight and an infrared spectrum that indicated a 1,2,4,6-tetrasubstituted set of aromatic rings. A product containing oxygen bridges (**43**, R = CH$_2$OCH$_2$), formed by intermolecular dehydration between a pair of molecules of **41**, would possess essentially the same molecular weight and infrared characteristics. In the same short paper Hunter and Turner reported a further extension of this process to the cyclodimerization of a pair of linear pentamers to produce, in poor but unstated yield, a cyclic decamer corresponding to **43** (R = CH$_2$) but containing two additional aromatic residues.

2.4.3 Chiral Calixarenes

One of the significant ways in which the calixarenes differ from the cyclodextrins is their lack of inherent chirality. Introduction of this feature into a

[60] R. F. Hunter and C. Turner, *Chem. Ind.*, **1957**, 72.

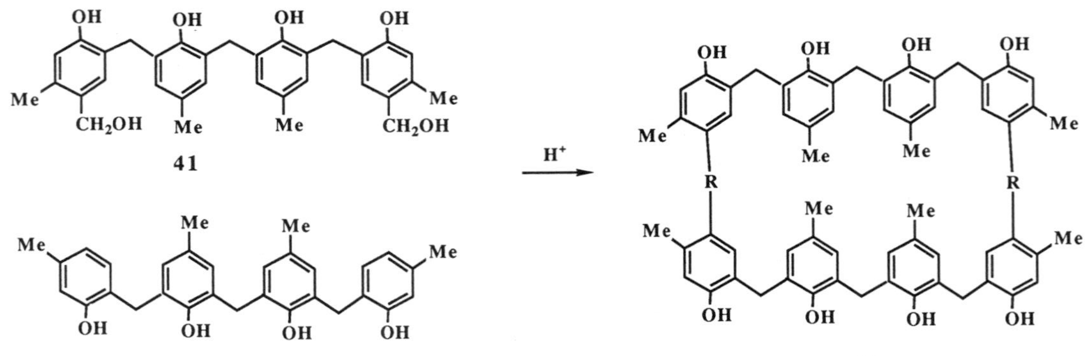

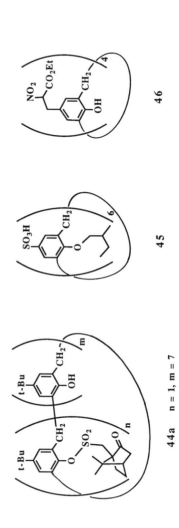

Figure 2.14 *Convergent stepwise synthesis of a calix[8]arene*[60]

44a n = 1, m = 7

44b n = 2, m = 6

45

46

Figure 2.15 *Calixarenes containing chiral centers in substituents at the 'lower rim'*[61,62] *and 'upper rim'*[63]

calixarene was first accomplished by Muthukrishnan and Gutsche[61] who prepared the mono- and di-camphorsulfonyl esters of *p-tert*-butyl-calix[8]arene (**44**) and studied their circular dichroic characteristics. More recently Shinkai and coworkers[62] have used the same 'lower rim functionalization approach' to introduce chirality by treating *p*-sulfonatocalix[6]arene with *S*-1-bromo-2-methylbutane to produce **45**, which was used for complexation studies to be discussed in Chapter 6. 'Upper rim' functionalization (see Chapter 5) has been used by Gutsche to make compound **46** containing four chiral centers.[63]

A more fundamental approach to introducing chirality into a calixarene is to incorporate it as an integral part of the calix. This was first accomplished, serendipitously, by No and Gutsche[46] who treated calixarene **47a** (synthesized by the method outlined in Figure 2.12) with acetyl chloride and AlCl₃ which yielded the symmetrical calixarene **47b**. When **47b** was subjected to a Baeyer–Villiger reaction with *m*-chloroperbenzoic acid only one of the acetyl groups underwent reaction to yield **47c** in which the **aabc** arrangement of the substituents on the 'upper rim' creates a chiral calix. A more systematic approach to the synthesis of inherently chiral calixarenes makes use of one or another of the stepwise syntheses by employing unsymmetrical building blocks. Böhmer and coworkers[49] achieved this in the reaction sequence described in Figure 2.7. All of the compounds prepared by route B shown in

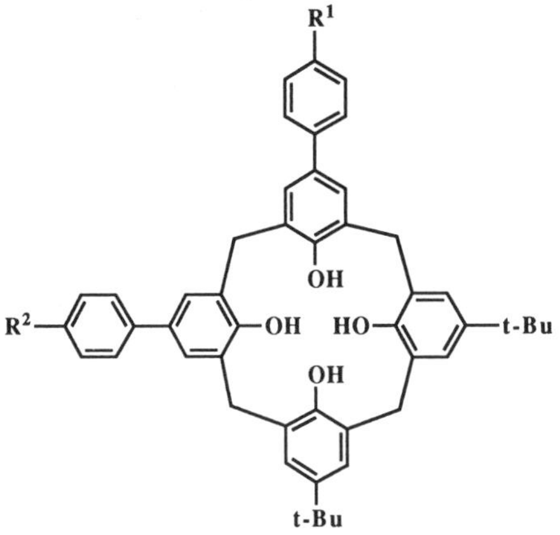

48

47a   R¹ = R² = H

47b   R¹ = R² = COCH₃

47c   R¹ = COCH₃, R² = OCOCH₃

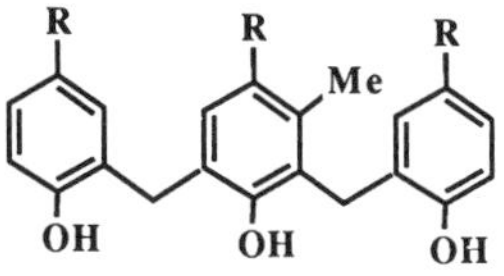

[61] R. Muthukrishnan and C. D. Gutsche, *J. Org. Chem.*, **1979**, *44*, 3962.
[62] S. Shinkai, T. Arimura, H. Satoh, and O. Manabe, *J. Chem. Soc., Chem. Commun.*, **1987**, 1495.
[63] C. D. Gutsche and K. C. Nam, *J. Am. Chem. Soc.*, **1988**, *110*, 6153.

Figure 2.7 contain an **aabc** sequence of aryl moieties and, therefore, are inherently chiral. A slightly different approach has been used by the group at Lyon[64] who used the Böhmer '3 + 1' method (see Figure 2.7) to join a *bis*-dibromomethyl monomer with a trimer (**48**) that is made unsymmetrical by the inclusion of a *m*-substituent. It should be pointed out that all of these inherently chiral calixarenes are incapable of resolution into enantiomers under ordinary conditions, because they are conformationally mobile. Of course, this is easy to change simply by converting them into esters or ethers, thereby freezing the compounds into one or another of the inflexible conformers. Conformational freezing, in fact, can convert an appropriately constructed calixarene that is inherently achiral in the cone conformation into one that is chiral in the partial cone conformation. For example, the bridged compound **70** shown in Figure 2.25 possesses chirality when it is placed in the partial cone conformation by appropriate functionalization at the OH groups (see Chapter 4).

2.5   Mechanisms of Calixarene-forming Reactions

2.5.1   Mechanism of the Base-induced Reaction

The mechanism, or perhaps more appropriately, 'pathway', of base-induced oligomerization of phenols and formaldehyde has been the subject of study for many decades. The first step is initiated by the formation of the phenoxide ion which, acting as a carbon nucleophile, effects a nucleophilic addition to the highly reactive carbonyl group of formaldehyde, *viz.*

Under mild conditions the reaction can be terminated at this point, and hydroxymethyl phenols can be isolated and characterized.[65] Under somewhat more strenuous conditions, however, the reaction proceeds further to give diarylmethyl compounds, presumably *via* a pathway that involves *o*-quinonemethide intermediates which react with phenolate ions in a Michael-like process, *viz.*

[64] H. Casabianca, J. Royer, A. Satrallah, A. Taty-C, and J. Vicens, *Tetrahedron Lett.*, **1987**, *28*, 6595.
[65] F. Ullman and K. Brittner, *Ber.*, **1909**, *42*, 2539.

For example, a mixture of *p-tert*-butylphenol and aqueous formaldehyde held at 50 °C for 4 days yields a mixture from which, *inter alia*, the *bis*-hydroxymethyl dimer can be isolated.[66] The possibility that *o*-quinone-methides might be intermediates in the formation of such oligomers was suggested as long ago as 1912[67] and was subsequently resurrected by Hultzsch,[68] v. Euler,[69] and others. Although a contrary view has been expressed[70] because of the very high temperatures that are required to transform methoxymethylphenols to quinonemethides,[71] it is known that reactions such as the oxy-Cope rearrangement proceed with far greater facility with the anions than with the corresponding neutral compounds.[72] It seems reasonable, therefore, to invoke the existence of quinonemethides under the conditions of the phenol–formaldehyde oligomerization reaction.

Another reaction of hydroxymethyl phenols that can occur instead of or concurrent with the formation of the diarylmethanes is dehydration to dibenzyl-type ethers, *viz.*

Thus, the mixtures from which calixarenes emerge contain diphenylmethane-type compounds and dibenzyl ether-type compounds in various extents of

[66] A. Zinke, R. Kretz, E. Leggewie, and K. Hössinger, *Monatsh. Chem.*, **1952**, *83*, 1213.
[67] A. Wohl and B. Mylo, *Ber.*, **1912**, *45*, 2046.
[68] K. Hultzsch, *Ber.*, **1942**, *75B*, 106.
[69] H. v. Euler, E. Adler, J. O. Cedwall, and O. Törngren, *Ark. Kemi, Mineral. Geol.*, **1941**, *15A*, No. 11, 1.
[70] R. Wegler and H. Herlinger, '*Methoden der Organischen Chemie* (Houben-Weyl), XIV/2, *Makromolecular Stoffe*', Thieme Verlag, Stuttgart, **1963**, p. 257.
[71] P. D. Gardner, H. Sarrafizadeh, and R. L. Brandon, *J. Am. Chem. Soc.*, **1959**, *81*, 5515.
[72] D. A. Evans and A. M. Golob, *J. Am. Chem. Soc.*, **1975**, *97*, 4765; R. W. Thies and E. P. Seitz, *J. Org. Chem.*, **1978**, *43*, 1050.

oligomerization. An HPLC analysis of the 'precursor' in a Modified Zinke–Cornforth Procedure (see Section 2.1) indicated at least three dozen noncyclic components to be present.[20] The precise mechanism for their transformation to calixarenes remains something of a mystery, although a speculative pathway has been suggested by Gutsche and coworkers[7] based on hydrogen bonding considerations and oligomer interconversion reactions.

That at least some of the components of the 'precursor' mixture are in equilibrium with one another prior to cyclization is suggested by the fact that the product mixture in a calixarene-forming reaction is much more dependent on the particular reaction conditions than on the starting material. For example, compounds **49—52** (R = H and CH_2OH) all give quite similar product mixtures under a particular set of conditions, *i.e.* Modified Zinke–Cornforth, Standard Petrolite, or Modified Petrolite Procedures.[7]

An important set of interconversions has been noted in the transformation of *p-tert*-butylcalix[8]arene or *p-tert*-butylcalix[6]arene to *p-tert*-butylcalix[4]arene in yields as high as 75%.[1,7] This process, which is induced by a combination of heat and base, has been characterized as 'molecular mitosis' by Gutsche who speculates that the cyclic octamer pinches together in the middle and then splits into a pair of cyclic tetramers, as depicted in Figure 2.16. Although there is some evidence that the transformation can also be induced from smaller to larger calixarenes, this remains to be firmly established.

When a mixture of *p-tert*-butylphenol and aqueous formaldehyde is treated with base, as described in Section 2.1.1 a 'precursor' is formed that undergoes transformation to calixarenes upon heating to higher temperatures — primarily to cyclic octamer in refluxing xylene or cyclic tetramer in refluxing diphenyl ether. In the latter case it has been observed that a precipitate forms well before the reflux temperature of diphenyl ether is reached, and isolation of the precipitate at this point shows it to be pure cyclic octamer. On the basis of the 'chemical mitosis experiment' described above, therefore, it is postulated that the cyclic octamer is an obligatory intermediate in the formation of cyclic tetramer. This turn of events is rationalized in terms of

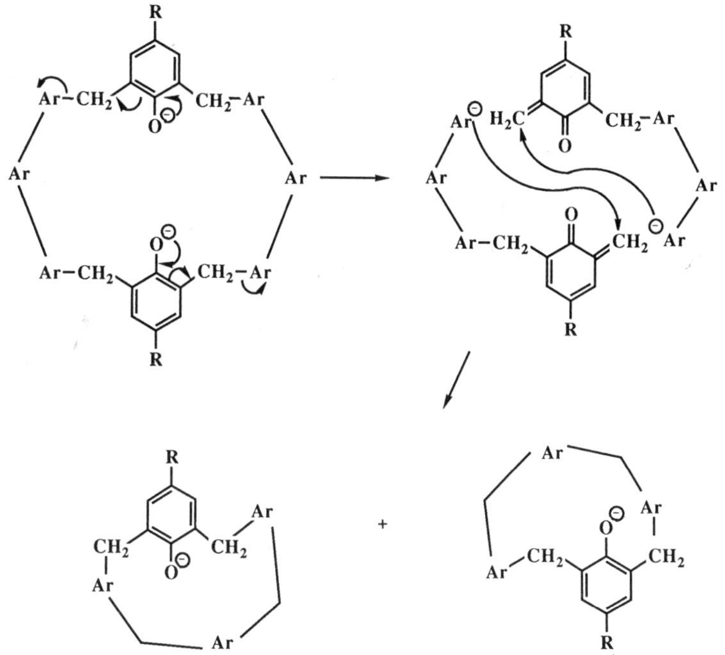

Figure 2.16 *Conversion of a calix[8]arene to a calix[4]arene: 'molecular mitosis'*[7]

hydrogen bonding in the cyclic as well as the linear oligomers. The presence of circularly hydrogen bonded arrays[73] in calixarenes is supported by infrared as well as *X*-ray crystallographic data which indicate that the calix[4]arenes exist in a 'cone' conformation and the calix[8]arenes in a 'pleated loop' conformation (see Chapter 4). Intramolecular hydrogen bonding also plays an important part in fixing the conformation of linear oligomers. *X*-Ray crystallography of a linear tetramer, for example, shows that it exists in a staggered conformation in the solid state,[74] intramolecular hydrogen bonding holding the phenyl groups in an almost planar, zig-zag array. If a linear tetramer in this conformation undergoes conversion into the cyclic tetramer a conformational inversion must occur that involves rotation around an axis passing through the center of the molecule. This produces a crescent-shaped species that has been called a 'pseudocalixarene'[75] (schematically represented in Figure 2.17), a transformation that must surmount not only the normal non-bonded barriers to rotation but also a certain amount of intramolecular hydrogen bond breaking. It is postulated[7] that a lower energy pathway involves intermolecular association of a pair of linear tetramers that have retained their 'zig-zag' conformation, forming a species that has been

[73] W. Saenger, C. Betzel, B. Hingerty, and G. M. Brown, *Angew. Chem., Int. Ed. Engl.*, **1983**, *22*, 883.

[74] E. Paulus and V. Böhmer, *Makromol. Chem.*, **1984**, *185*, 1921.

[75] B. Dhawan and C. D. Gutsche, *J. Org. Chem.*, **1983**, *48*, 1536.

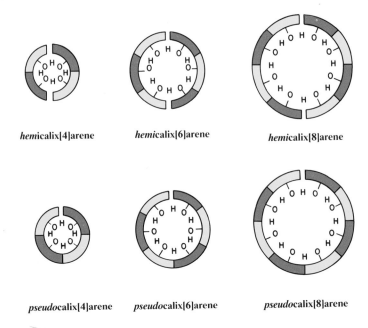

*hem*icalix[4]arene   *hem*icalix[6]arene   *hem*icalix[8]arene

*pseudo*calix[4]arene   *pseudo*calix[6]arene   *pseudo*calix[8]arene

Figure 2.17   *Schematic representations of hemicalixarenes and pseudocalixarenes*

called a 'hemicalixarene'[75] (schematically represented in Figure 2.17). Effective circular hydrogen bonding should exist in a *hem*icalix[8]arene, enhancing its stability. Subsequent conversion to a calix[8]arene entails no conformational changes nor any significant disruption of the hydrogen bond array of the 'pleated loop'. Thus, the process of cyclic tetramer formation is thought to be that depicted in Figure 2.18.

p-tert-Butyldihomooxacalix[4]arene (**53**) has been isolated from reactions of *p-tert*-butylphenol and formaldehyde[76] carried out under the Petrolite conditions, suggesting that it might be a precursor to the calix[4]arene (**54**), as shown in Figure 2.19. However, when **53** is heated separately in diphenyl ether in the presence of base, little or no **54** can be detected. The pathways of decomposition of dibenzyl-type ethers have been studied in connection with the curing of Bakelite[77-79] but are still unclear.[80]

The mechanism of formation of the calix[6]arenes is even more conjectural. Since Lin has shown[81] that *p-tert*-butylcalix[8]arene can be con-

[76] C. D. Gutsche, R. Muthukrishnan, and K. H. No, *Tetrahedron Lett.*, **1979**, 2213.
[77] K. Hultzsch, *Ber.*, **1941**, *74*, 898, 1533.
[78] H. v. Euler, E. Adler, and J. O. Cedwall, *Ark Kemi, Mineral. Geol.*, **1941**, *14A*, No. 14, 1; H. v. Euler, E. Adler, J. O. Cedwall, and O. Törngren, *ibid.*, **1941**, *15A*, No. 11, 1.
[79] A. Zinke, E. Ziegler, and I. Hontschik, *Monatsh. Chem.*, **1948**, *48*, 317.
[80] In addition to water and formaldehyde, other products such as benzaldehydes have been shown to be formed when dibenzyl-type ethers are heated above 150 °C (Ref. 79 and E. Ziegler and I. Hontschik, *ibid.*, **1948**, *78*, 325).
[81] L-G. Lin, PhD Thesis, Washington University, St Louis, **1984**.

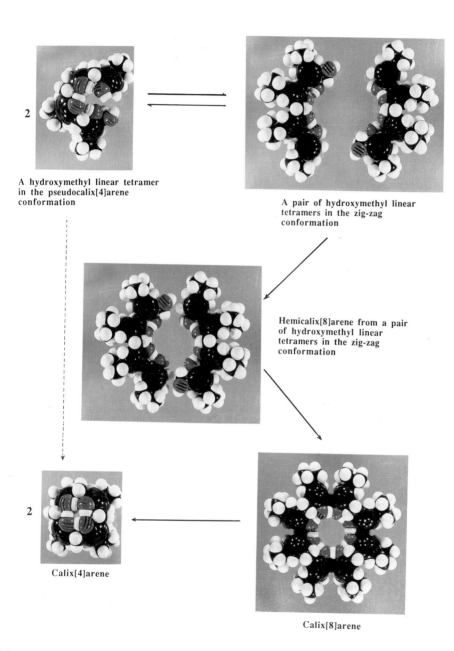

A hydroxymethyl linear tetramer
in the pseudocalix[4]arene
conformation

A pair of hydroxymethyl linear
tetramers in the zig-zag
conformation

Hemicalix[8]arene from a pair
of hydroxymethyl linear
tetramers in the zig-zag
conformation

Calix[4]arene

Calix[8]arene

Figure 2.18 *Conversion of a linear tetramer into a calix[8]arene and a calix[4]arene, as depicted with space-filling molecular models[7]*

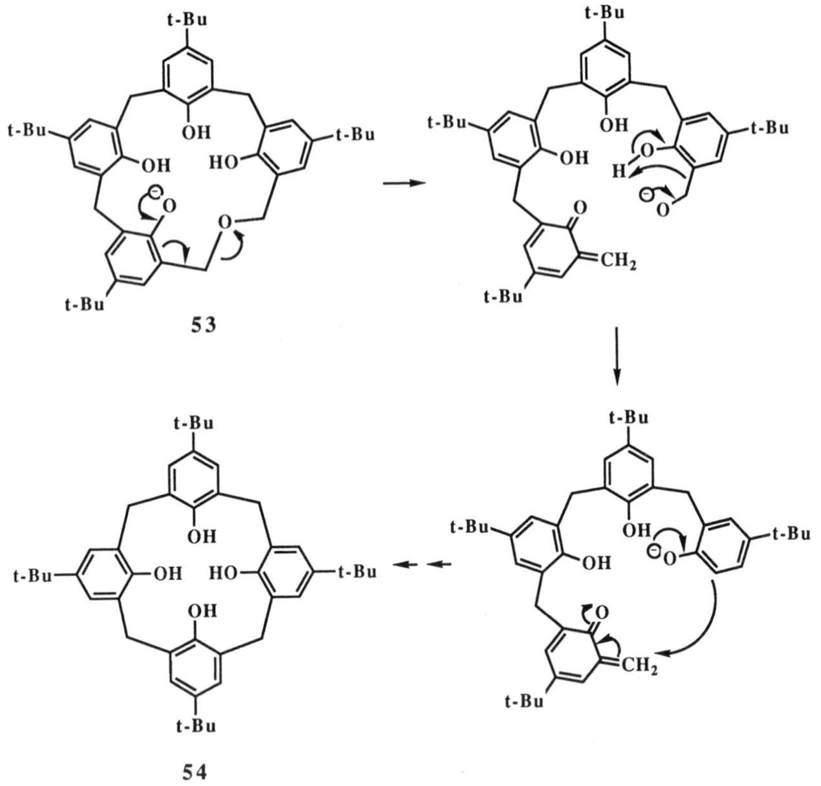

Figure 2.19 *Pathway for the base-induced conversion of dihomooxacalix[4]arenes into calix[4]arenes*[7]

verted either into the calix[4]arene, as described above, or into the calix[6]arene, depending on the amount of base that is used, a pathway for the formation of the calix[6]arenes analogous to that postulated for the calix[4]arenes provides one possibility, *viz.* calix[8]arene as the obligatory intermediate. Alternatively, a calix[6]arene might arise *via* a hemi-calix[6]arene by a pathway analogous to that postulated for the formation of the calix[8]arene from a hemicalix[8]arene, as schematically illustrated in Figure 2.20. Space-filling molecular models indicate, however, that if a calix[6]arene is to exist in a 'pleated loop' conformation it probably can have no more than four hydrogens in the circular array of oxygens, *i.e.* that it must exist as a dianion. The requirement for a 10-fold increase in the amount of base to shift the product from the calix[8]arene to the calix[6]arene in the Petrolite Procedure accords with this postulate. Also, the somewhat greater yield of calix[6]arene in the presence of KOH or RbOH as compared with CsOH, NaOH, or LiOH suggests that a template effect is occurring, a pheno-menon that has been well documented in crown ether chemistry.[82]

---

[82] G. W. Gokel and S. H. Korzeniowski, '*Macrocyclic Polyether Syntheses*', Springer, Berlin, **1982**, Chapter 2.

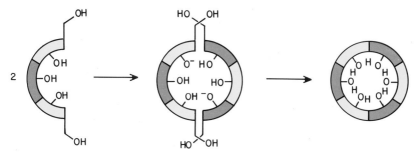

Figure 2.20 *Conversion of a linear trimer into a calix[6]arene* via *the dianion of a* hemi*calix[6]arene*[7]

A mechanism similar to that portrayed in Figure 2.18 can also be invoked to explain the origins of calixarenes of other sizes. Thus, a hemicalix[5]arene formed by association of a linear dimer and trimer would yield a calix[5]arene; a hemicalix[7]arene formed by the association of a linear trimer and tetramer would yield a calix[7]arene, *etc.* The amounts of the various hemicalixarenes that are formed might depend not only on the relative concentrations of the linear oligomers present but also on the effectiveness of hydrogen bonding in the hemicalixarenes. As discussed in Chapter 4, calixarenes containing an even number of aromatic rings appear to be more conformationally stable than those with an odd number, probably reflecting their relative strengths of intramolecular hydrogen bonding. The general tendency for the even numbered calixarenes to be formed in higher yields finds explanation, therefore, in the assumption that the hydrogen bonding in the calixarenes parallels that in the hemicalixarenes.

The outcome of the base-induced condensation of formaldehyde with *p*-substituted phenols is remarkably sensitive to the *p*-substituent. Few other reactions among the thousands now known to chemists appear to be as fastidious in their structural requirements. A detailed study[7] of calixarene formation from *p*-isopropyl-, *p-tert*-butyl-, *p-tert*-pentyl, and *p*-(1,1,3,3-tetramethylbutyl)phenol led to the conclusion that the composition of the product mixture must be the result of the interplay of a variety of factors including the hydrogen bonding, the temperature, and the template effects already discussed. Even among these four phenols there are significant variations in outcome, with *p-tert*-butylphenol giving the best yields and the most tractable products. More profound structural variation in the *para*-substituent results in the formation of far less tractable products, and even *p*-cresol yields a mixture that appears to be calixarene-like in composition but that has not yet yielded to characterization. Most phenols, particularly those substituted with electron-withdrawing substituents, either fail to react or give products other than calixarenes. Why the *p*-substituent exerts so powerful an influence in calixarene formation is not yet known, but one must conclude that in the cases of the *p*-alkylphenols it probably arises from steric rather than electronic factors. The substituent effect appears to be more critical to the formation of calix[4]arenes than the larger cyclooligomers. *p*-Phenylphenol, for example, reacts with formaldehyde to yield cyclic hexamer, heptamer, and octamer but no cyclic tetramer. When the 'molecular mitosis' experiment described above for the conversion of *p-tert*-butylcalix[8]arene into *p-tert*-butylcalix[4]arene is attempted with *p*-phenylcalix[8]arene no reaction occurs. This suggests that the failure to detect the cyclic tetramer from direct calixarene-forming reactions is the result of the inability of this system to undergo the cycloreversion reaction, and it provides further circumstantial evidence in support of the mechanism portrayed in Figure 2.18. The ease with which a cyclic tetramer forms, either in the cycloreversion experiment or in a direct calixarene-producing reaction, is perhaps related to the degree to which the *p*-substituents fill the cavity created in the 'cone' conformer. Four *tert*-butyl groups nicely cover the cavity, as do four *p-tert*-pentyl or four 1,1,3,3-tetramethylbutyl groups. But, four isopropyl groups do this somewhat less well, and four methyl groups or four phenyl groups rather poorly. Perhaps the 'molecular mitosis' process is particularly sensitive to the way in which the *p*-substituents interact with one another as the molecule pinches at the midpoint prior to undergoing cycloreversion.

To summarize the present postulates concerning base-induced calixarene formation,[7] the conversion of hydroxymethylated linear oligomers into cyclic oligomers involves the formation of hemicalixarenes in which the forces of intermolecular and intramolecular hydrogen bonding play a crucial role. By extrusion of water and formaldehyde the hemicalixarene is transformed to the calixarene. In the presence of very small amounts of base, with solvents boiling below 180 °C, and with phenols in which the *p*-alkyl substituents cannot fold over the top of the cyclic tetramer the major product is the cyclic octamer, which precipitates from solution and is assumed to be a product of *kinetic control*. In the presence of larger amounts of base and one of the larger cations the major product is the cyclic hexamer which is viewed as a product of kinetic control and/or a *template effect*. Under higher temperature conditions the cyclic tetramer is the major product in those cases where the *p*-substituent can fold over the top of the cone conformer, and it is viewed as the product of *thermodynamic control*.

2.5.2   Mechanism of the Acid-catalyzed Reaction

The acid-catalyzed condensation of resorcinol with aldehydes is most logically interpreted in terms of cationic intermediates (**55, 57**) and electrophilic aromatic substitution reactions to form **56, 58** and **59** as portrayed in Figure 2.21. It is not known whether the cyclic tetramer forms by cyclodimerization of a pair of hydroxymethylated dimers derived from **58** or from a simple cyclization of a hydroxymethyl linear tetramer. As in the base-induced process, it is not clear what the driving forces are for cyclization. Although the eight extraannular OH groups of resorcinol-derived calixarenes cannot engage in circular hydrogen bonding, pairwise hydrogen bonding is

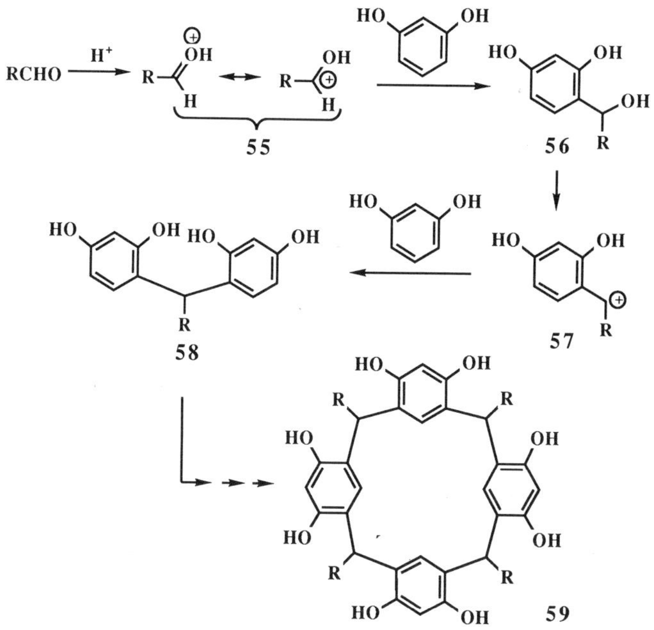

Figure 2.21 *Pathway for the acid-catalyzed condensation of resorcinols and aldehydes*

possible and may play a dominant role in organizing the system for cyclization even under acidic conditions.[25]

2.5.3 Mechanism of the Thermally-induced Reaction

The thermally-induced condensations of phenols and formaldehyde perhaps are unexpected because of the long-standing emphasis on base and acid catalysis. Whether these are truly neutral reactions or whether traces of acid or base are present to act as catalysts cannot yet be certain. The formation of a calixarene under neutral conditions (see Figure 2.4) is interesting and may possibly be the result of intermolecular hydrogen bonding which facilitates the formation of a hemicalix[4]arene and holds the reactants in an array that is conducive to calix[4]arene formation.

2.6 Homocalixarenes, Oxacalixarenes, and Homooxacalixarenes

Although this book focuses primarily on compounds possessing $[1_n]$meta-cyclophane structures, there are a number of compounds that are closely enough related to permit them to be viewed as relatives of the calixarenes. Included in this group are the homocalixarenes, the oxacalixarenes, and the homooxacalixarenes.

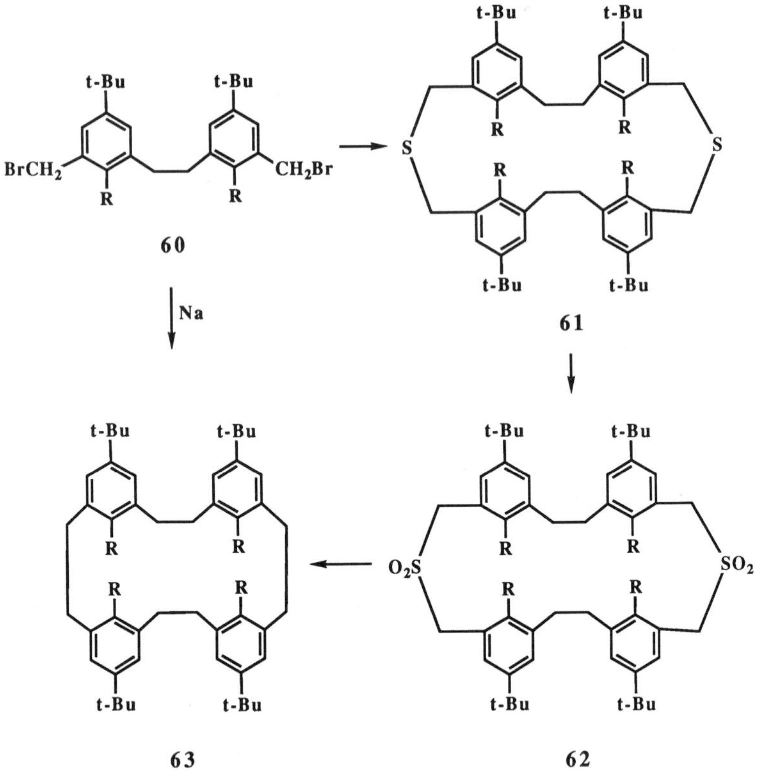

Figure 2.22 *Synthesis of tetrahomocalixarenes*[83, 84]

2.6.1 Homocalixarenes

Tashiro and coworkers have reported the synthesis of several 2,8,14,20-tetrahomocalix[4]arenes, initially as side products in low yield from the treatment of **60** with sodium[83] and more recently by employing the oft-used sulfone extrusion method (**61** → **62** → **63**) for macrocyclic synthesis,[84] as illustrated in Figure 2.22. Among the compounds reported is **63** (R = OH), obtained by BBr$_3$-induced demethylation of the corresponding tetramethyl ether.

2.6.2 Oxacalixarenes

[1$_n$]Metacyclophanes containing hetero atoms rather than methylene groups as bridges are known. For example, the 2,8,14,20-tetraoxacalix[4]arene **64** (Y = oxygen) and the 2,14-diaza-8,20-dioxacalix[4]arene **64** (Y = nitrogen)

[83] M. Tashiro and T. Yamato, *J. Org. Chem.*, **1981**, *46*, 1543.
[84] M. Tashiro and A. Tsuge, 12th International Symposium on Macrocyclic Chemistry, Hiroshima, **1987**, Abstracts, p. 95.

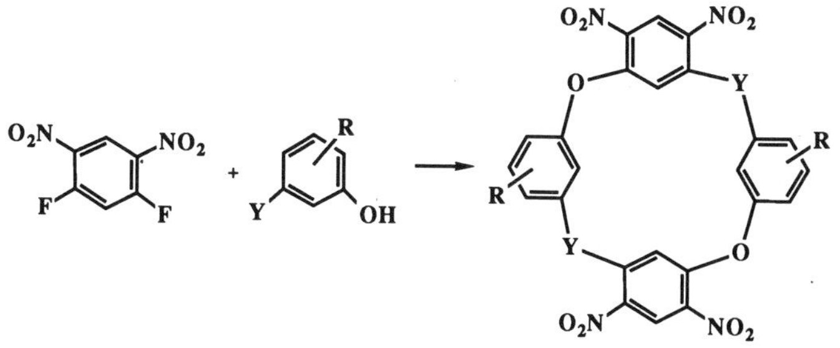

64

Figure 2.23 *Synthesis of oxacalixarenes and oxaazacalixarenes*[85-88]

have been prepared by condensing *m*-dichloroarenes[85,86] or *m*-difluoro-arenes[87] with resorcinols or *m*-aminophenols, as illustrated in Figure 2.23. The tetrathio analog of **64** has also been prepared.[88] Like the carbo-calix[4]arenes, these compounds possess nonplanar structures and have melting points in the 350—400 °C range.

2.6.3 Homooxacalixarenes

In the Standard Petrolite Procedure for calixarene synthesis one of the components of the product mixture is *p-tert*-butyldihomooxacalix[4]arene (**66**),[75,89] present in large amounts in the early phases of the reaction but giving way to calixarenes as the reaction progresses. That it is a true obligatory intermediate in calixarene formation, however, seems doubtful (see Section 2.5). Even in mixtures where it is present in significant amount it is difficult to isolate and purify, and a useful alternative to its preparation employs the dehydration of the bis-hydroxymethyl linear tetramer **65**, as shown in Figure 2.24. Cyclization occurs in almost quantitative yield, and this has been used to advantage in the preparation of the bridged dihomooxa-calix[4]arene[90] **70** from the linear tetramer **69**, as shown in Figure 2.25. A variety of other bis-hydroxymethylphenols have been dehydrated to cyclic ethers, including 2,6-bis-hydroxymethylphenols (**71**) which yield products that are assumed to be cyclic trimers (**72**)[75,91] on the basis of mass spectral molecular weight determinations. Dehydrations of the bis-hydroxymethyl linear dimers **73** (X = CH$_2$) yield tetrahomodioxacalix[4]arenes **74** (X = CH$_2$), and in similar fashion the analogous dibenzyl ethers **73** (X = CH$_2$OCH$_2$) yield **74** (X = CH$_2$OCH$_2$), as shown in Figure 2.26. In a somewhat related reaction,

[85] N. Sommer and H. A. Staab, *Tetrahedron Lett.*, **1966**, 2837.
[86] E. E. Gilbert, *J. Heterocyclic Chem.*, **1974**, *11*, 899.
[87] P. A. Lehman, *Tetrahedron*, **1974**, *30*, 727.
[88] G. Montaudo, F. Bottino, and E. Trivellone, *J. Org. Chem.*, **1972**, *37*, 504.
[89] T. Tanno and Y. Mukoyama, *Netsu. Kokasei Jushi*, **1981**, *2*, 132 (*Chem. Abstr.*, 96: 52791k).
[90] C. D. Gutsche, P. K. Sujeeth, and D. W. Bailey, unpublished experiments.
[91] Y. Mukoyama and T. Tanno, *Org. Coating Plastics Chem.*, **1979**, *40*, 894.

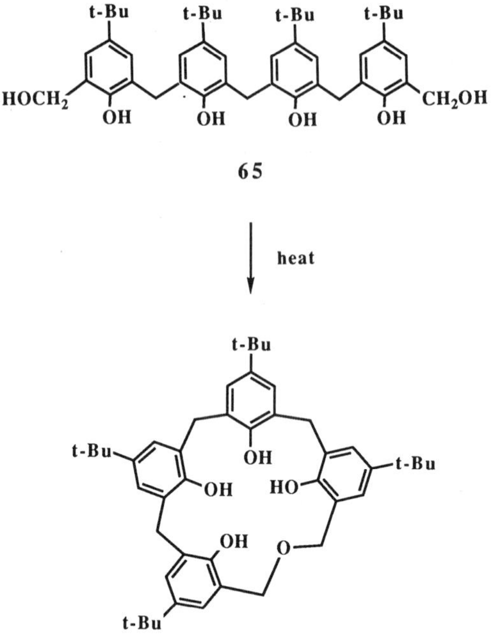

Figure 2.24   *Synthesis of* p-tert-*butyldihomooxacalix[4]arene*[75]

the bis-hydroxymethyl compound derived from *m*-xylene (**75**) undergoes acid-catalyzed dehydration to yield cyclic hexamers and cyclic octamers (**76**, *n* = 6 and 8).[92]

2.7   Heterocalixarenes

Somewhat further afield, but yet closely enough related to be considered at least first cousins of the calixarenes are the products obtained from condensations of furans, thiophenes, or pyrroles with aldehydes or ketones. Furan reacts with aldehydes and ketones under acidic conditions to give cyclic tetramers of structure **77**.[93–96] The doubling of the yields in the presence of lithium perchlorate[94] suggested that a template effect might be operating, but later experiments[95] showed that the higher yields correlate better with the acidity of the reaction medium than with the identity of the metal ion.[96] Thiophene and pyrrole undergo similar acid-catalyzed reactions

[92] M. P. Weaver and C. Y. Meyers, *Tetrahedron Lett.*, **1959**, 7.
[93] W. H. Brown and B. J. Hutchinson, *Can. J. Chem.*, **1978**, *56*, 617 and preceeding papers back to W. H. Brown and H. Sawatzky, *ibid.*, **1956**, *34*, 1147.
[94] M. Chastrette and F. Chastrette, *J. Chem. Soc., Chem. Commun.*, **1973**, 534.
[95] M. de S. Healy and A. J. Rest, *J. Chem. Soc., Chem. Commun.*, **1981**, 149.
[96] A. G. S. Högberg and M. Weber, *Acta Chem. Scand. B*, **1983**, *37*, 55.

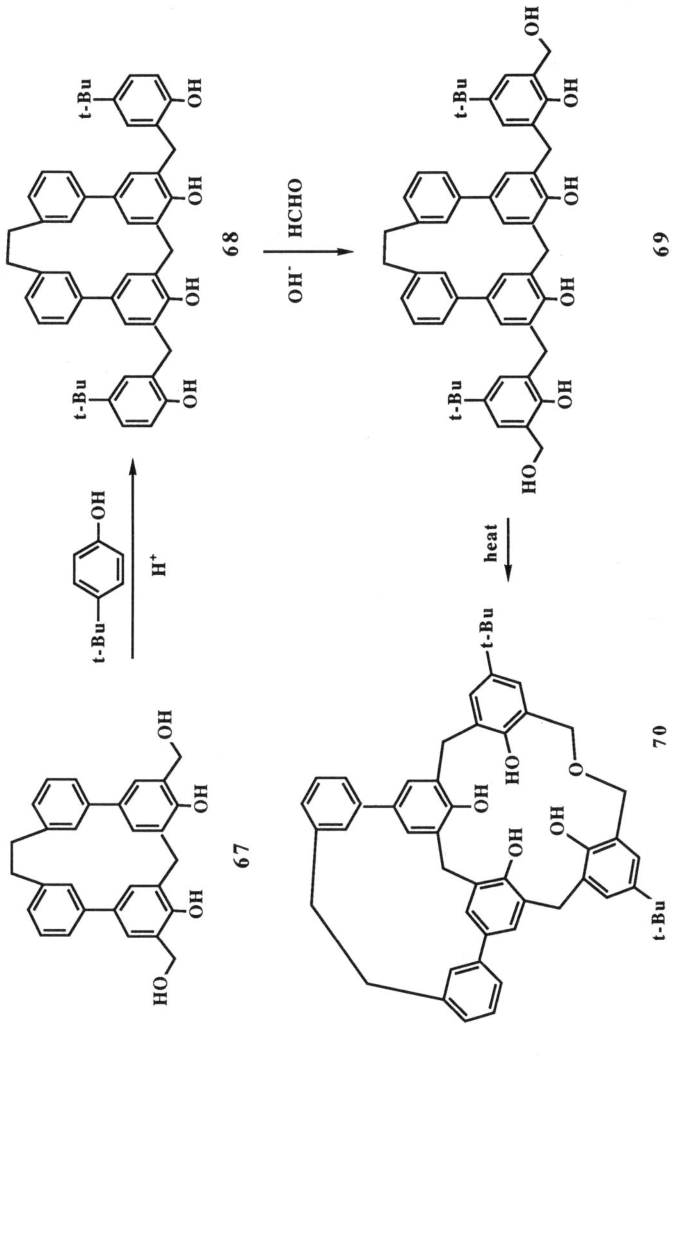

Figure 2.25 *Synthesis of a bridged dihomooxacalix[4]arene*[90]

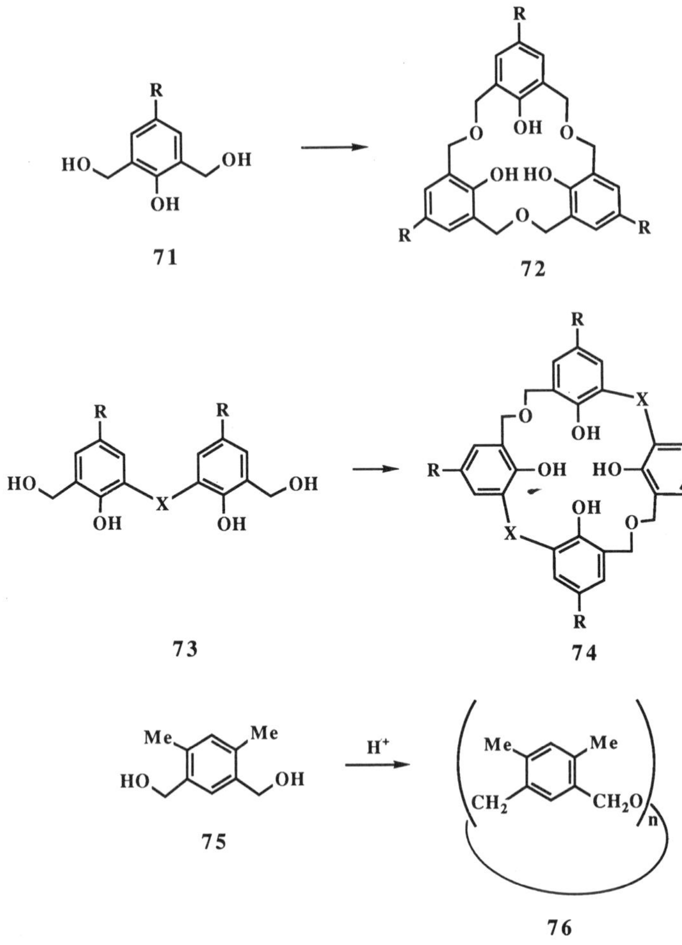

Figure 2.26 *Cyclodehydration of bis-hydroxymethylarenes*[75, 91, 92]

with acetone to yield cyclic tetramers **78**[97] and **79**,[98] as illustrated in Figure 2.27. Aromatic aldehydes and pyrroles yield cyclic tetramers,[99] but in these cases the initially-formed products undergo oxidation to planar porphyrin systems that are no longer sufficiently calixarene-like to be considered a close relation.

A pyridinocalix[4]arene has been prepared by George Newkome and coworkers[100] by the dimerization of the bis-(2-bromopyridyl) ketal **80** to

[97] M. Ahmed and O. Meth-Cohn, *Tetrahedron Lett.*, **1969**, 1493; *J. Chem. Soc., Ser. C*, **1971**, 2104.

[98] P. Rothemund and C. L. Gage, *J. Am. Chem. Soc.*, **1955**, 77, 3340.

[99] J. S. Lindsey, I. C. Schreiman, H. C. Hsu, P. C. Kearney, and A. M. Marguerettaz, *J. Org. Chem.*, **1987**, 52, 827.

[100] G. R., Newkome, Y. J. Joo, and F. R. Fronczek, *J. Chem. Soc., Chem. Commun.*, **1987**, 854.

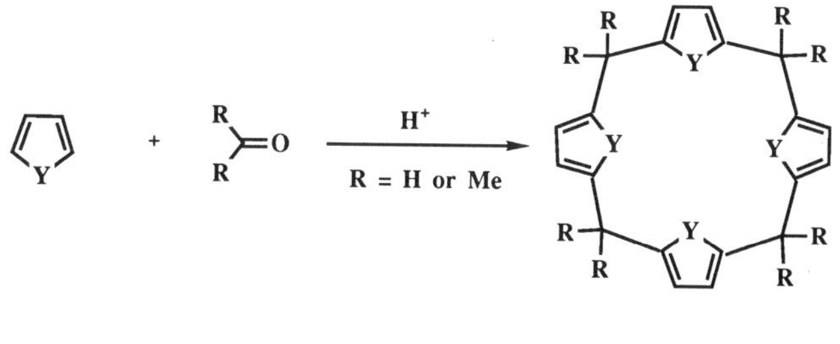

77 (Y = O)

78 (Y = S)

79 (Y = N)

Figure 2.27 *One-step, acid-catalyzed synthesis of heterocalixarenes[93–99]*

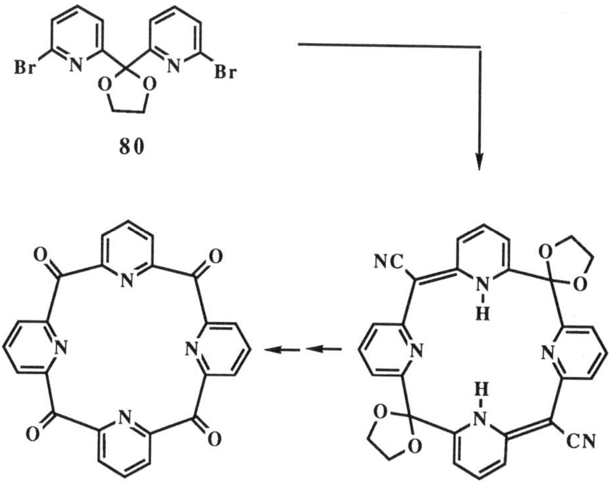

82   **81**

Figure 2.28 *Stepwise synthesis of a pyridinocalix[4]arene[100]*

yield **81** from which the CN groups were removed, hydrolysis of the ketals effected, and oxidation of the remaining methylene groups carried out with SeO$_2$ to yield **82**, as illustrated in Figure 2.28. Still another variety of nitrogen-containing calixarene ring system has been reported by Ian Sutherland and coworkers[101] who treated the cyclic imide **83** with NaH and bis-1,3-bromomethylbenzene (**84**) and obtained a 7% yield of the bis-tetrahydropyrimidone calixarene **85**, as illustrated in Figure 2.29.

[101] J. A. E. Pratt, I. O. Sutherland, and R. F. Newton, *J. Chem. Soc., Perkin Trans. 1*, **1988**, 13.

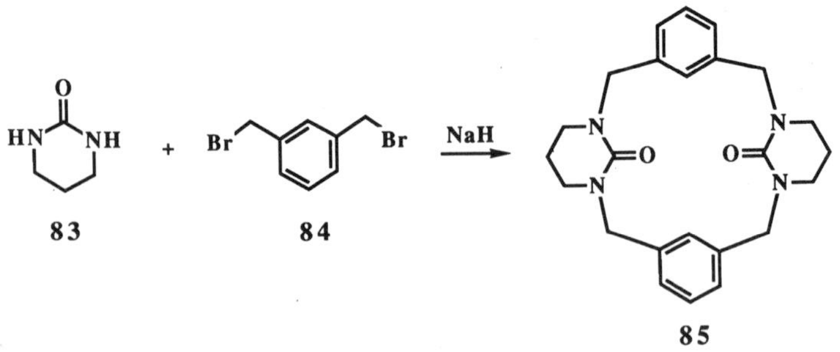

Figure 2.29 *Synthesis of a bis-tetrahydropyrimidone calixarene[101]*

2.8 Concluding Remarks

In the first chapters of this book the early history of phenol–formaldehyde chemistry is portrayed, with particular reference to the events that laid the basis of the experiments that were initiated in the 1970's in three different laboratories. Chapter 1 discussed the work from the St Louis laboratory which focused on the unraveling of the older literature and the introduction and development of one-step methods for calixarene synthesis. The present chapter has discussed the contributions from the Mainz laboratory where Professors Kämmerer and Böhmer exploited previously introduced linear stepwise methods and invented new convergent methods for calixarene synthesis. The next chapter discusses methods for establishing the structure of calixarenes, bringing to the fore the research group at Parma.

e Baskets: The
zation and Properties
nes

. .....ays happens that when a man seizes upon a neglected and important idea, people inflamed with the same notion crop up all around'

Mark Twain, *Life on the Mississippi*

In 1975 Rocco Ungaro of the University of Parma spent a year at the State University of New York in Syracuse as a research associate with Professor Johannes Smid and was introduced to the new area of crown ether chemistry. Upon returning to his home laboratory in Parma, Ungaro sought a way of combine this new field with his earlier interest in phenol–formaldehyde chemistry. Already aware of the papers by Zinke and Cornforth, he and his colleague Andrea Pochini decided that the 'cyclic tetramers' should make

The Parma Group Giovanni Andreetti, Rocco Ungaro, and Andrea Pochini

67

good 'ordered matrices' for attaching crown ether-type ligands for cation binding. Repeating Cornforth's experiments, they were immediately confronted with the complications discussed in Chapter 1. After a year of frustration in trying to deal with the mixtures that are produced by the Zinke-Cornforth procedure, they decided in 1977 to enlist the services of another colleague, Giovanni Andreetti, hoping to take advantage of his expertise as an *X*-ray crystallographer. The collaboration proved fruitful, and by the end of the year an *X*-ray structure of Cornforth's lower-melting compound (LBC) had been solved. However, thinking that no one else in the world could possibly be working on the Zinke–Cornforth compounds, they were induced to publish this result only when the 1978 paper by Gutsche and Muthukrishnan[1] revealed the fallacy of this assumption. Starting in 1979 with the publication of the crystal structure of the Cornforth LBC compound,[2] the group at the University of Parma has continued its unusually productive collaboration and has contributed some of the most important advances in the field of calixarene chemistry.

3.1   *X*-Ray Crystallography: The Ultimate Proof of Structure

3.1.1   Phenol-derived Calixarenes

Chemistry, though dealing with materials that generally can be seen and felt, relies on a theory of matter that postulates particles whose existence must be inferred from indirect observation. The creation of the atomic and molecular structural theory has been rightly hailed as one of the supreme creations of the human intellect, the fact of which our cultural elite needs gentle reminding.[3] With the exception of the recently developed scanning tunneling microscope which gives a direct display of atoms and molecules in a few selected cases,[4] the closest the chemist can come to actually seeing molecules is through the use of *X*-ray crystallography. The way that this technique has revolutionized our knowledge of the large biomolecules of nature is a well-known and still exciting story. Its impact on the chemistry of smaller molecules, though less dramatic, is no less important, and in the case of the calixarenes, as with many other molecules, *X*-ray crystallography provides the conclusive proof of structure.

It is worth noting, however, that the gross structures of all of the calixarenes, from the calix[4] to the calix[8]arene, were correctly inferred from existing evidence prior to confirmation by *X*-ray crystallography. For example, the

---

[1] C. D. Gutsche and R. Muthukrishnan, *J. Org. Chem.*, **1978**, *43*, 4905.

[2] G. D. Andreetti, R. Ungaro, and A. Pochini, *J. Chem. Soc., Chem. Commun.*, **1979**, 1005.

[3] 'Science as well as art illuminates man's view of himself and his relation to others', David Baltimore in *Daedalus*, **1988**, *117 (3)*, 335.

[4] For example, *cf. Chem Eng. News*, **1988**, August 1, p. 5 for a picture of benzene.

high-melting product from *p-tert*-butylphenol and formaldehyde was first established as the cyclic octamer[5] by its reaction with a limiting amount of 2,4-dinitrochlorobenzene which yielded a compound with an ArO:OH ratio of 1:7. Subsequently, osmometric and mass spectral data reinforced this conclusion. The success of such conventional analytical techniques notwithstanding, the application of *X*-ray crystallography to calixarene chemistry has had a powerful impact in providing the definitive proof of structure.

The first example of an *X*-ray crystallographic structure of a calixarene was the one reported in 1979 by Andreetti, Ungaro, and Pochini[2] of the Cornforth LBC compound, followed four years later by a structure of Cornforth's LOC compound.[6] Both were shown to possess the cyclic tetrameric structure, as pictured in Figure 3.1 for the LBC compound. An especially interesting feature of this structure is the toluene molecule *inside* the cyclic tetramer, the result of a complexation phenomenon that is discussed in Chapter 6. *X*-Ray crystallographic confirmation for the cyclic pentamer structure has been obtained by Ninagawa[7] for the *p-tert*-butyl compound and by the Parma group[8] for calix[5]arene itself, as shown in Figure 3.1. For a number of years *p-tert*-butylcalix[6]arene posed the typical *X*-ray crystallographic problem of yielding beautiful glistening crystals that were too small to be easily handled. Andreetti and coworkers[9] overcame this hurdle and have reported the *X*-ray structure for the cyclic hexamer shown in Figure 3.1. *p-tert*-Butylcalix[8]arene presented another often encountered problem, yielding glistening needles that quickly lose their morphology and crumble to a powder. The St Louis group[10] with the help of the *X*-ray crystallographer Alexander Karaulov of the University of York, circumvented this problem by crystallizing the compound from pyridine and maintaining the crystal in a pyridine-saturated atmosphere to obtain the structure shown in Figure 3.1. To date there has been no report of an *X*-ray crystallographic structure of a calix[7]arene or the recently isolated *p-tert*-butylcalix[9]arene.

X-Ray crystallographic structures have been obtained for numerous calixarene derivatives in addition to the parent calixarenes. Typical representatives include the tetraacetate of *p-tert*-butylcalix[4]arene,[11] the hexamethyl ether of *p*-allylcalix[6]arene,[12] and the octaacetate of *p-tert*-butylcalix[8]arene,[13] as pictured in Figure 3.2.

[5] R. Muthukrishnan and C. D. Gutsche, *J. Org. Chem.*, **1979**, *44*, 3962.

[6] G. D. Andreetti, A. Pochini, and R. Ungaro, *J. Chem. Soc., Perkin Trans. 2*, **1983**, 1773.

[7] A. Ninagawa, personal communication.

[8] M. Coruzzi, G. D. Andreetti, V. Bocchi, A. Pochini, and R. Ungaro, *J. Chem. Soc., Perkin Trans. 2*, **1982**, 1133.

[9] G. D. Andreetti, G. Calestani, F. Ugozzoli, A. Arduini, E. Ghidini, A. Pochini, and R. Ungaro, *J. Inclusion Phenom.*, **1987**, *5*, 123.

[10] C. D. Gutsche, A. E. Gutsche, and A. I. Karaulov, *J. Inclusion Phenom.*, **1985**, *3*, 447.

[11] C. Rizzoli, G. D. Andreetti, R. Ungaro, and A. Pochini, *J. Mol. Struct.*, **1982**, *82*, 133.

[12] L.-g. Lin, G. F. Stanley, and C. D. Gutsche, unpublished work (see L.-g. Lin, PhD Thesis, Washington University, St Louis, **1984**, p. 41).

[13] G. D. Andreetti, R. Ungaro, and A. Pochini, *J. Chem. Soc., Chem. Commun.*, **1981**, 533.

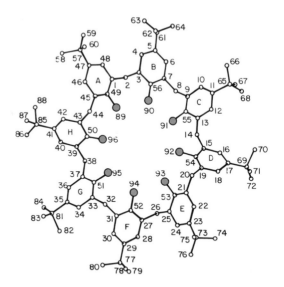

p-tert-Butylcalix[4]arene

Calix[5]arene

p-tert-Butylcalix[6]arene

p-tert-Butylcalix[8]arene

Figure 3.1 X-*Ray crystallographic structures of* p-tert-*butylcalix[4]arene,* calix[5]arene, p-tert-*butylcalix[6]arene, and* p-tert-*butyl-* .calix[8]arene[2, 8–10]

The X-ray crystallographic structures shown in Figures 3.1 and 3.2 provide conclusive evidence for the monocyclic character of the calixarenes, laying to rest such possibilities as the cyclic octamer being a pair of cyclic tetramers tethered to one another in some fashion. In addition to its power in establishing the gross structures of the calixarenes, X-ray crystallography has given great insight into the subtle features of shape and complex formation, topics that are discussed in Chapters 4 and 6, respectively. X-Ray crystallography, in fact, now pervades all of supramolecular chemistry and has become an indispensable tool of investigation.

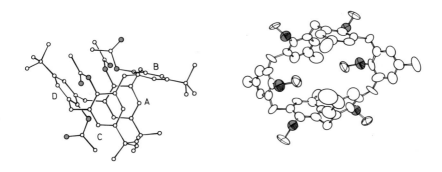

Tetra-acetate of *p-tert*-butylcalix[4]arene   Hexa-methyl ether of *p*-allylcalix[6]arene

Octa-acetate of *p-tert*-butylcalix[8]arene

Figure 3.2   *X-Ray crystallographic structures of calixarene derivatives*[11−13]

3.1.2   Resorcinol-derived Calixarenes

The products of acid-catalyzed condensation of resorcinol and aldehydes, first postulated by Niederl in 1940 to have a cyclic tetrameric structure,[14] became the object of detailed study in the 1960's in the laboratories of Holger Erdtman at the Karolinska Institute in Stockholm. Erdtman and coworkers carried out reactions with resorcinol and various aldehydes[15] and obtained crystals suitable for *X*-ray analysis from the octa-butyrate of C-*p*-bromophenylcalix[4]resorcinarene. An *X*-ray determination on this material[16] firmly established the validity of Niederl's structure and showed the compound to have an all-*cis* set of configurations with the four *p*-bromophenyl groups all oriented in the same direction, as illustrated in Figure 3.3. The structure of a second isomer obtained from this same reaction was determined in a similar fashion *via* its octaacetate[17] and shown to have the *cis,trans,cis,trans* configuration. C-Methylcalix[4]resorcinarene, bridged between the neighboring OH groups with dimethylsilyl moieties, has also been shown to possess the all-*cis* configuration.[18] In contrast to the phenol-derived calixarenes, where several of the parent compounds have yielded to *X*-ray

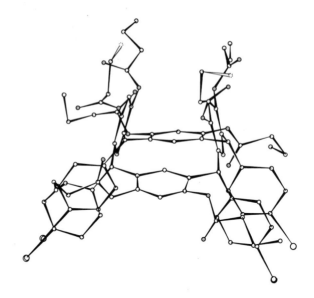

Figure 3.3   X-*Ray crystallographic structure of* C-p-*bromophenyl-calix[4]resorcin-arene in the all-*cis* configuration*[16]

[14] J. B. Niederl and H. J. Vogel, *J. Am. Chem. Soc.*, **1940**, *62*, 2512.
[15] H. Erdtman, F. Haglid, and R. Ryhage, *Acta Chem. Scand.*, **1964**, *18*, 1249; H. Erdtman, S. Högberg, S. Abrahamsson, and B. Nilsson, *Tetrahedron Lett.*, **1968**, 1679.
[16] B. Nilsson, *Acta Chem. Scand.*, **1968**, *22*, 732.
[17] K. J. Palmer, R. Y. Wong, L. Jurd, and K. Stevens, *Acta Crystallogr., Sect B*, **1976**, *32*, 847.
[18] D. J. Cram, K. D. Stewart, I. Goldberg, and K. N. Trueblood, *J. Am. Chem. Soc.*, **1985**, *107*, 2574.

analysis, the resorcinol-derived calixarenes have all required conversion into derivatives to obtain suitable crystals.

3.2 Proof of Structure by Chemical Interconversion

One of the classical methods for proof of structure involves conversion of the compound of unknown structure to one of known structure by means of unequivocal reactions. At least two examples exist in the field of calixarenes, the first being the proof of structure of *p*-(1,1,3,3-tetramethyl-butyl)calix[8]arene (Cornforth's HOC compound). Using reactions that are discussed in Chapter 5, the Parma scientists[19] removed the *tert*-butyl groups of *p-tert*-butylcalix[8]arene and the 1,1,3-3-tetramethylbutyl groups of the HOC compound and obtained the same product from both reactions, *viz.* calix[8]arene, as shown in Figure 3.4. Since the structure of the *tert*-butyl compound is known from *X*-ray crystallography, the structures of the compounds related to it in this fashion are confidently established.

p-tert-Butylcalix[4]arene also provides a link between the one-step and multi-step syntheses, as illustrated in Figure 3.5. Using reactions that are discussed in Chapter 5, Gutsche and Nam[20] converted *p-tert*-butyl-calix[4]arene (which can be prepared by the one-step process) to *p*-methyl-

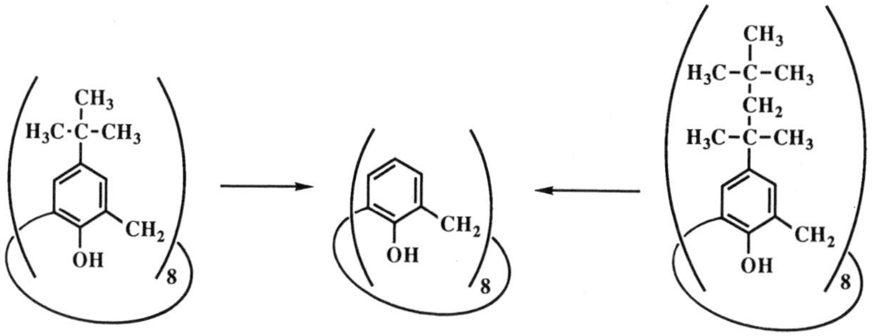

Figure 3.4   *Proof of structure of Cornforth's HOC compound[19]*

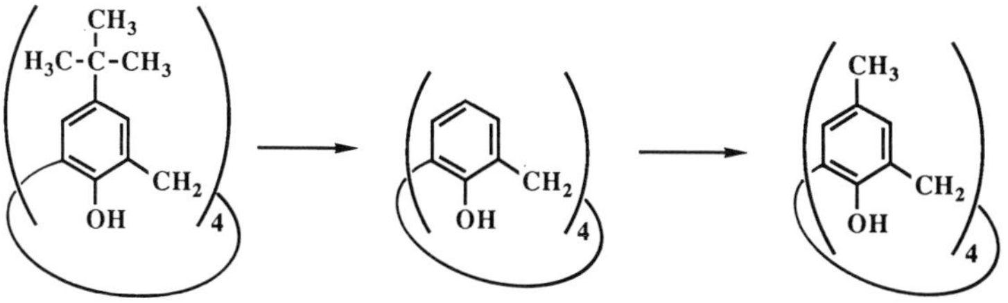

Figure 3.5   *Connection between the one-step and multi-step synthesis of* p-methyl-calix[4]arene[20]

[19] V. Bocchi, D. Foina, A. Pochini, R. Ungaro and G. D. Andreetti, *Tetrahedron*, **1982**, *38*, 373.
[20] C. D. Gutsche and K. C. Nam, *J. Am. Chem. Soc.*, **1988**, *110*, 6153.

calix[4]arene (which can be prepared by the Hayes and Hunter multi-step process).

3.3 Physical Properties of Calixarenes

3.3.1 Melting Points

To jump from X-ray crystallography to melting points is to go from one end of the scale of analytical sophistication to the other, from one of the most recent of techniques to one of the oldest. Melting points, though, have played a significant part in calixarene chemistry, for it was their unusually high values that first attracted Zinke's attention. It is now recognized that this is a characteristic property of many calixarenes, especially those with free hydroxyl groups where melting points usually are above 250 °C. For example, p-tert-butylcalix[4]arene melts at 342—344 °C, p-tert-butylcalix[6]arene at 380—381 °C, p-tert-butylcalix[8]arene at 411—412 °C, p-phenylcalix[4]arene at 407—409 °C, and p-phenylcalix[8]arene above 450 °C. The substituent in the p-position of the calixarene ring, however, can have a pronounced effect, and it is reported[21] that some of the calix[6]arenes produced from p-n-alkyl-phenols (ranging from p-n-octyl to p-n-octadecyl) have melting points as low as 110 °C. No experimental details are furnished in this paper however. The low melting points reported by Moshfegh et al.[22] for p-halocalix[4]arenes must also be questioned (see Chapter 2, p. 42—44). A well-documented illustration of the subtle effect that substituents can exert is provided by Böhmer's compounds[23] 18g, 18h, and 18i in Figure 2.7 which have melting points of 185—190 °C, 270 °C, and 368 °C, respectively, a remarkable range for isomeric compounds that differ only in the arrangement of identical groups around the 'upper rim' of the calix.

The melting point and its range provide a quick indication of the purity of the p-tert-butylcalixarenes. Melting point diagrams[24] for two calixarene mixtures are shown in Figure 3.6 where it is noted that even a small amount of the second component quite significantly lowers the melting point and increases the melting range.

Alterations in the calixarene structure via derivatization, of course, can greatly affect the melting point, and esters and ethers of calixarenes generally melt lower than the parent compound. For example, the tetramethyl and tetrabenzyl ethers of p-tert-butylcalix[4]arene melt at 226—228 and 230—231 °C, respectively. But, there are exceptions; the trimethylsilyl ether of this same calixarene melts at 411—412 °C, and the tetraacetate melts at 383—386 °C.

---

[21] Z. Asfari and J. Vicens, Tetrahedron Lett., 1988, 29, 2659.
[22] A. A. Moshfegh, R. Badri, M. Hojjatie, M. Kaviani, B. Naderi, A. H. Nazmi, M. Ramezanian, B. Roozpeikar, and G. H. Hakimelahi, Helv. Chim. Acta, 1982, 65, 1221; A. A. Moshfegh, B. Mazandarani, A. Nahid, and G. H. Hakimelahi, ibid., 1982, 65, 1229; A. A. Moshfegh, E. Beladi, L. Radnia, A. S. Hosseini, S. Tofigh, and G. H. Hakimelahi, ibid., 1982, 65, 1264.
[23] V. Böhmer, F. Marschollek, and L. Zetta, J. Org. Chem., 1987, 52, 3200.
[24] We are indebted to Mr. Matthew Wolf for obtaining these data.

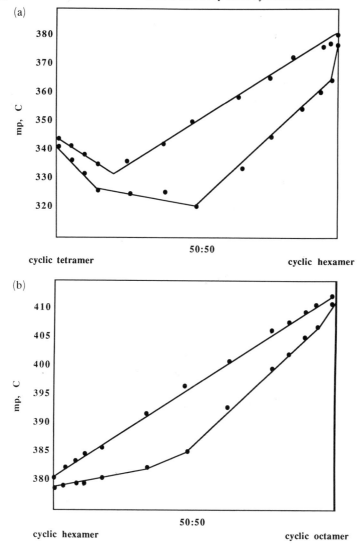

Figure 3.6 *Melting point diagrams for (a) mixtures of* p-tert-butylcalix[4]arene and p-tert-butylcalix[6]arene *and (b) mixtures of* p-tert-butylcalix[6]arene *and* p-tert-butylcalix[8]arene. *The lower values represent the onset of melting and the higher values the completion of the melting[24]*

3.3.2 Solubilities

Another characteristic feature of the calixarenes that caught Zinke's attention is their insolubility in water and aqueous base and their low solubility in organic solvents. In fact, it is this feature that prevented Zinke from obtaining what he considered to be a satisfactory molecular weight (see Chapter 1), and it is this feature that makes some of the calixarenes so difficult to isolate, purify, and characterize. Fortunately, most calixarenes have sufficient solubility in chloroform, pyridine, or carbon disulfide to permit spectral determinations to be made, although one exception is the very high-melting calixarene from p-adamantylphenol[25] that is so insoluble that it has defied adequate characterization. Just as the p-alkyl groups in a calixarene affect its melting point so too they affect solubility, and it is no surprise that long-chain alkyl groups in the p-positions increase the solubility in organic solvents. For example, in a study of the temperature-dependent ^{1}H NMR characteristics of the calixarenes, Munch[26] chose the p-octyl rather than the p-*tert*-butyl compound because of its greater solubility in chloroform. Conversion of calixarenes into esters or ethers generally increases the solubility in organic solvents, and this difference can sometimes be used to separate the components of a mixture. For example, p-phenylcalix[6]arene and p-phenylcalix[8]arene have been separated from a reaction mixture and from one another by employing a continuous extraction procedure,[27] and No et al.[28] have used the solubility of the reactant (calix[4]arene tetraacetate) in hot benzene to separate it from the product (p-acetylcalix[4]arene) which is insoluble in this medium.

It has been of particular interest to confer water solubility on the calixarenes. This was first achieved early in 1984 by Ungaro and coworkers[29] who prepared the tetracarboxymethyl ether (1) of p-*tert*-butylcalix[4]arene, soluble to the extent of 5×10^{-3} to 5×10^{-4} M depending on the accompanying cation. Calixarenes containing 4—8 aromatic rings and carrying carboxyl groups attached to the p-positions (2,3) have been prepared by Gutsche and coworkers[30,31] and found to have solubilities of *ca.* 10^{-3} M in dilute aqueous base. Considerably more soluble than the carboxycalixarenes are the sulfonated calixarenes (4) which were prepared by Shinkai and coworkers[32,33] late in 1984. They reported the p-sulfonato calix[4], [6], and [8] arenes and found them to have solubilities at least as great as 0.1 M. A number of aminocalixarenes (5,6) have been synthesized by the St Louis group[20,31] and are moderately soluble in dilute aqueous acid. The special interest in the water soluble calixarenes stems from their ability to interact with ions and molecules and to perform catalysis, topics that are discussed in detail in later chapters of this book.

Although the p-alkylphenol-derived calixarenes have little solubility in aqueous base, the resorcinol-derived calix[4]arenes form tetraanions[34] in

[25] S.-I. Chen, Ph.D. Dissertation, Washington University, St Louis, **1985**, p. 46–7.

[26] J. H. Munch, *Makromol. Chem.*, **1977**, *178*, 69.

[27] P. F. Pagoria, PhD. Thesis, Washington University, St Louis, **1984**, p. 123.

[28] K. No, Y. Noh, and Y. Kim, *Bull. Korean Chem. Soc.*, **1986**, *7*, 442.

[29] A. Arduini, A. Pochini, S. Reverberi, and R. Ungaro, *J. Chem. Soc., Chem Commun.*, **1984**, 981.

[30] C. D. Gutsche and P. F. Pagoria, *J. Org. Chem.*, **1985**, *50*, 5795.

[31] C. D. Gutsche and I. Alam, *Tetrahedron*, **1988**, 4689.

[32] S. Shinkai, S. Mori, T. Tsubaki, T. Sone, and O. Manabe, *Tetrahedron Lett.*, **1984**, *25*, 5315.

[33] S. Shinkai, K. Araki, T. Tsubaki, T. Arimura, and O. Manabe, *J. Chem. Soc., Perkin Trans. 1*, **1987**, 2297.

[34] H.-J. Schneider, D. Güttes, and U. Schneider, *Angew. Chem., Int. Ed. Engl.*, **1986**, *25*, 647.

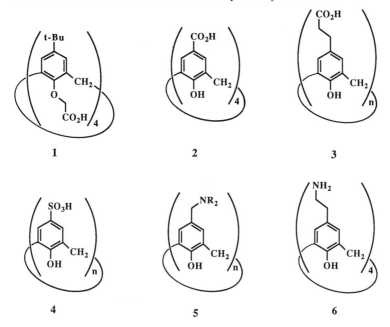

Figure 3.7 *Water-soluble calixarenes*

0.01 M NaOH that are sufficiently soluble to allow complexation studies to be carried out, as discussed in Chapter 6. The esters of the resorcinol-derived calixarenes, of course, are insoluble in aqueous media but soluble in organic solvents.

3.3.3 pK_a Values

Early attempts to obtain dissociation constants for the OH groups of the *p-tert*-butylcalixarenes were frustrated by the low solubility of these compounds. Success was first achieved by Böhmer and coworkers[35] with the mono-nitrocalixarenes **7a** and **7b** whose pK_1 values in 1:1 water:methanol were obtained by measuring the UV absorption as a function of the pH, using 10 cm cuvettes. The values obtained were 6.0 for **7a** and 4.3 for **7b**, showing that these calixarenes are somewhat more acidic than *p*-nitrophenol. The surprising 1.7 pK unit difference in the pK values of the two compounds, which differ only in a single *p*-alkyl substituent, is attributed to subtle conformational effects arising from the relative sizes of the methyl and *tert*-butyl groups.

With the advent of the very water-soluble *p*-sulfonatocalixarenes and the moderately water-soluble *p*-nitrocalixarenes from Shinkai's laboratory[36] the measurements of pK_a values have yielded data for all four of the OH groups

[35] V. Böhmer, E. Schade, and W. Vogt, *Makromol. Chem., Rapid Commun.*, **1984**, 5, 221.
[36] S. Shinkai, K. Araki, H. Koreishi, T. Tsubaki, and O. Manabe, *Chem. Lett.*, **1986**, 1351.

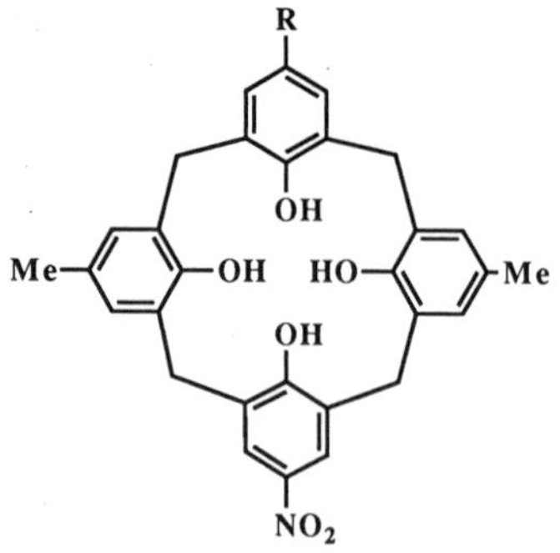

7a R = Me

7b R = t-Bu

in the calix[4]arenes, as shown in Table 3.1. Calixarenes **8a** and **8b** are both seen to be far stronger acids than their monomeric counterparts, possessing pK_1 values of less than 1 and less than 0, respectively. Thus, they are more acidic than trichloroacetic acid and trifluoroacetic acid. Also interesting are

Table 3.1   *pK Values for* p-*sulfonatocalix[4]arene and* p-*nitrocalix[4]arene*[36]

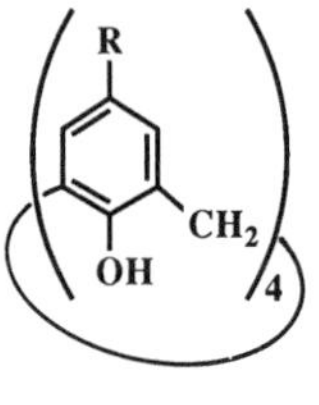

Compound	Solvent	pK_1	pK_2	pK_3	pK_4
8a SO₃H	H$_2$O	<1	3.0±0.4	4.0±0.4	>11
	H$_2$O	8.9±0.1			
8b NO₂	H$_2$O–THF (7:3)	<0	10.3±0.3	13±1	
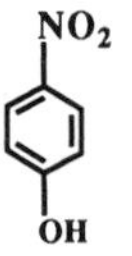	H$_2$O–THF (7:3)	7.1±0.1			

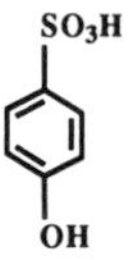

8a R=SO₃Na
8b R=NO₂

the differences between **8a** and **8b** with respect to their pK_2 and pK_3 values and the spread between their pK_1 and pK_4 values.

3.4   Spectral Properties of Calixarenes

3.4.1   Infrared Spectra

One of the particularly distinctive features of the calixarenes is the unusually low frequency at which the stretching vibration of the OH groups occurs in the infrared, ranging from *ca.* 3150 cm^{-1} for the cyclic tetramer to *ca.* 3300 cm^{-1} for the cyclic pentamer, with the other calixarenes falling in between these extremes. This is the result of the especially strong intramolecular hydrogen bonding that exists in these molecules, a feature that was discussed in the previous chapter in connection with the pathway for calixarene formation (see Section 2.5.1). 'Circular hydrogen bonding'[37] undoubtedly is responsible for some of the unique features of the calixarenes, and a recent study by Tobiason and coworkers[38] using FTIR confirms the intramolecular character of the bonding and shows that it is strongest for the cyclic tetramer and weakest for the cyclic pentamer. Low stretching frequencies have been observed for linear phenol–formaldehyde oligomers as well,[39] presumably because of the formation of pseudocalixarenes and/or hemicalixarenes (see Figure 2.17). Subtle changes in the shape of the hydrogen bond array can result from the imposition of 'upper rim' bridges as shown by the arrichoarenes (see Figure 2.9) where the strength of the hydrogen bonding increases (*i.e.* decreasing v_{OH} values) as the bridge gets longer and the calix becomes more flexible.

In the 'fingerprint' region the calixarenes all look rather similar to one another, especially between 1500 and 900 wavenumbers. In the 500–900 cm^{-1} region, however, the patterns vary to some extent, as shown in the spectra displayed in Figure 3.8. For example, absorptions that appear to be characteristic of the individual calixarenes are found at 693 and 571 cm^{-1} in the cyclic pentamer, 762 cm^{-1} in the cyclic hexamer, and 796 cm^{-1} in the cyclic heptamer. The cyclic octamer is distinguished by less well-resolved absorptions in the 500–600 cm^{-1} region. A band near 400 cm^{-1} is stated to be useful in differentiating cyclic tetramer from cyclic hexamer and cyclic octamer.[40] The alkyl ethers of the calix[4]arenes and calix[6]arenes have strong absorptions at 850 cm^{-1} and 810 cm^{-1}, respectively.

3.4.2 Ultraviolet Spectra

The linear as well as the cyclic oligomers have a pair of absorption maxima at 280 and 288 nm in the ultraviolet region. In the latter case the ratio of the absorptions at these two wavelengths is a function of the size of the ring, ranging from 1.3 for the cyclic tetramers to 0.75 for the cyclic octamers, as shown in Table 3.2.[41–45] Although reactions today are generally monitored by means of TLC, HPLC, or ^{1}H NMR measurements, UV has been used on occasion for this purpose. Cornforth and coworkers[46] followed the course of

[37] W. Saenger, C. Betzel, B. Hingerty, and G. M. Brown, *Angew. Chem., Int. Ed. Engl.*, **1983**, *22*, 883.
[38] S. W. Keller, G. M. Schuster, and F. L. Tobiason, *Polym. Mater. Sci. Eng.*, **1987**, *57*, 906.
[39] T. Cairns and G. Eglinton, *Nature*, **1962**, *196*, 535.
[40] Professor R. T. Hawkins, Brigham Young University, personal communcation.
[41] H. Kämmerer and G. Happel, *Makromol. Chem.*, **1978**, *179*, 1199.
[42] Dr. F. J. Ludwig, Petrolite Corporation, unpublished observations.
[43] H. Kämmerer, G. Happel, and B. Mathiasch, *Makromol. Chem.*, **1981**, *182*, 1685.
[44] H. Kämmerer and G. Happel, *Monatsh. Chem.*, **1981**, *112*, 759.
[45] H. Kämmerer and G. Happel, *Makromol. Chem.*, **1980**, *181*, 2049.
[46] J. W. Cornforth, E. D. Morgan, K. T. Potts, and R. J. W. Rees, *Tetrahedron*, **1973**, *29*, 1659.

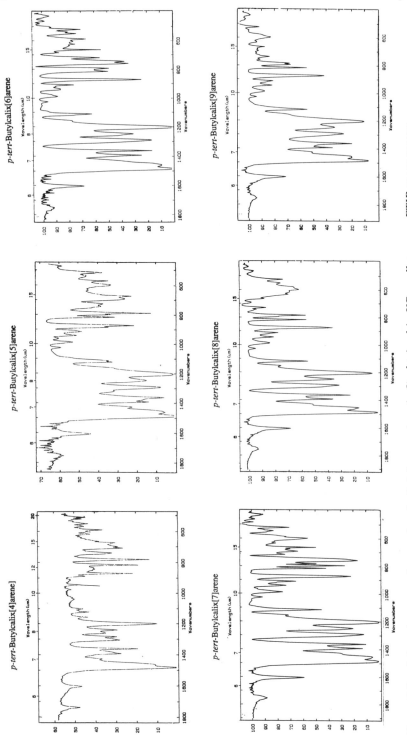

Figure 3.8 *Infrared spectra of p-tert-butylcalix[n]arenes (n = 4—9) obtained in KBr pellets on an FTIR spectrometer*

Table 3.2   *Absorptivities* $(\varepsilon_{max}, mol^{-1}\ cm^{-1})$ *of calixarenes in the indicated solvent at* 280 nm *and* 288 nm

R Group	Ring size	280 ± 1 nm	288 ± 1 nm	Solvent	Ref
All *tert*-butyl	4	9,800	7,700	CHCl$_3$	43
All methyl	4	10,500	8,300	Dioxane	42
Me & *tert*-butyl	5	14,030	14,380	Dioxane	44
All *tert*-butyl	6	15,500	17,040	CHCl$_3$	43
Me & *tert*-butyl	6	17,210	17,600	Dioxane	45
All *tert*-butyl	7	18,200	20,900	CHCl$_3$	43
Me & *tert*-butyl	7	19,800	20,900	Dioxane	46
All *tert*-butyl	8	23,100	32,000	CHCl$_3$	43

the oxyalkylation of calixarenes by noting the disappearance of the absorption at 300 nm (characteristic of the free phenol) and the growth of an absorption at 270—280 nm arising from the ether.

3.4.3   NMR Spectra

When calixarenes re-emerged in the 1970's NMR spectroscopy had become one of the most important analytical tools available to the organic chemist, and the utility of this powerful technique is clearly indicated by observing the difference in the ^{13}C NMR spectra of the linear and cyclic oligomers, as shown in Figure 1.4. Because of the symmetry of the cyclooligomers their spectra, regardless of ring size, are less complex than those of their acyclic counterparts and in the case of the *p-tert*-butyl compounds consists of four resonance lines arising from the aromatic carbons, one from the methylene carbons, and two from the *tert*-butyl carbons. While this simple spectrum is not an infallible proof of structure of a calixarene, it provides a valuable and easily obtained early clue.

The ^{1}H NMR spectra of the symmetrically-substituted calixarenes are similarly uncomplicated, that for *p-tert*-butylcalix[4]arene at room temperature being illustrated in Figure 3.9. The resonances from the aromatic protons, the *tert*-butyl protons, and the hydroxyl protons are singlets, and that from the CH$_2$ protons is a pair of doublets which is generally the most useful aspect of the spectrum. As Kämmerer first showed,[47] the spectrum of a calix[4]arene in chloroform at *ca.* 20 °C displays a pair of doublets which collapse to a singlet when the temperature is raised to *ca.* 60 °C. On the basis of NOE experiments and the shift induced by pyridine as a solvent, Ungaro and coworkers[29] assigned the higher field doublet to the equatorial protons (closer to the aromatic rings) and the lower field doublet to the axial protons (closer to the hydroxyl groups). Gutsche and coworkers[48] demonstrated that

---

[47] H. Kämmerer, G. Happel, and F. Caesar, *Makromol. Chem.*, **1972**, *162*, 179.
[48] C. D. Gutsche, B. Dhawan, K. H. No, and R. Muthukrishnan, *J. Am. Chem. Soc.*, **1981**, *103*, 3782; *ibid.*, **1984**, *106*, 1891.

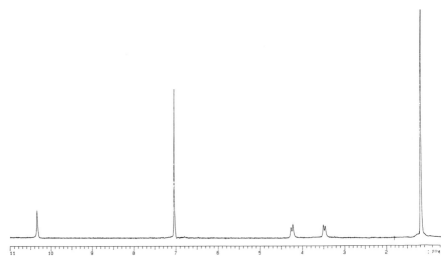

Figure 3.9   *¹H NMR spectrum of* p-tert-*butylcalix[4]arene*

calix[8]arenes behave in a comparable fashion but that calix[6]arenes show only a singlet resonance at room temperature and must be taken to a lower temperature before a splitting pattern emerges. The calix[4] and calix[8]arenes can be differentiated from one another[49] by the fact that in pyridine solution the calix[4]arene retains a pair of doublets at low temperature (*e.g. ca.* 0 °C) whereas the calix[8]arene shows only a singlet even at temperatures as low as − 90 °C. The structure of *p*-isopropylcalix[8]arene, for example, was established on the basis of observations of this sort.[50] The calix[5]arenes, calix[7]arenes, and calix[9]arenes have been much less studied than the even-numbered members because of their difficult accessibility by either the one-step or the multi-step syntheses. The odd-numbered calixarenes are closer to the calix[6]arene than to the calix[4] or calix[8]arenes with respect to their ¹H NMR behavior in nonpolar solvents such as chloroform.

The position of the singlet arising from the OH groups varies with the ring size of the calixarene and does not correlate particularly well with the strength of hydrogen bonding as judged by the ν_{OH} frequencies in the IR (see Table 4.5). An 'upper rim' bridged calixarene (see Figure 2.9) carrying an 8-carbon spanner has been found by Böhmer and coworkers[51] to exhibit two OH resonances at − 60 °C (δ 9.26 and δ 9.14) that collapse to a singlet (δ 9.05) at 25 °C. The analogous compound carrying a 10-carbon spanner, however, fails to show this behavior because of the greater flexibility of the

[49] C. D. Gutsche and L. J. Bauer, *Tetrahedron Lett.*, **1981**, *22*, 4763.
[50] J. Vicens, T. Pilot, D. Gamet, R. Lamartine, and R. Perrin, *C. R. Acad. Sci. S., Paris*, **1986**, *t302* II, 15.
[51] V. Böhmer, H. Goldmann, R. Kaptein, and L. Zetta, *J. Chem. Soc., Chem. Commun.*, **1987**, 1358.

calix. The OH resonance from *p-tert*-butylcalix[9]arene which is a singlet at room temperature, gives rise to no fewer than nine lines at − 30 °C.[52]

3.4.4 Mass Spectra

Molecular weight determinations of calixarenes have played a crucial role ever since they were first reported by Zinke in 1941. Restricted in the early days to cryoscopic or ebulloscopic methods, the lack of solubility of the calix-arenes posed serious problems. As discussed in Chapter 1, this resulted in some missed opportunities at correct structure assignments. With the advent of good osmometers fairly accurate molecular weights could be determined with small amounts of quite insoluble materials, and this technique provided the St Louis group with valuable data in the early 1970's. Today, the most generally useful and powerful tool for obtaining molecular weights is the mass spectrometer, particularly with the introduction of fast atom bombardment and other new techniques.

The first mass spectral determination of a calixarene was carried out in 1964 by Erdtman and Ryhage[53] at the Karolinska Institute in Stockholm, one of the early centers for mass spectrometry. They used a calixarene derived from resorcinol and acetaldehyde which was converted into the octamethyl ether with diazomethane. Observation of a molecular ion signal at *m/e* 656 proved the presence of four aromatic residues and provided confirmation for the cyclic tetrameric structure that Niederl had assigned 25 years earlier. When the reinvestigation of the phenol-derived calixarenes began in the 1970's mass spectrometry also played a useful role, although sometimes with misleading information. A case in point is the mass spectrum of *p-tert*-butyl-calix[8]arene which shows a strong signal at *m/e* 648, in exact agreement with a cyclic tetramer. In fact, for several years this datum was accepted as support for a calix[4]arene structure of the cyclooligomer produced by the Standard Petrolite Procedure. The appearance of weak signals with *m/e* greater than 648, however, was disturbing and suggested that something was amiss. Con-tinuing attention to this puzzle eventually led to the preparation of a tri-methylsilyl derivative which showed a very strong signal at *m/e* 1872 (corresponding to a cyclic octamer) along with a signal at *m/e* 936 (cor-responding to a cyclic tetramer) that was only slightly weaker than that of the parent ion. Whether the latter signal arises from a dication or from the cleavage of the cyclic octamer in the mass spectrometer (*e.g.* 'molecular mitosis' — see Figure 2.16) is not known, but it seems probable that it arises from cleavage. Mass spectral data have been reported for what were thought to be *p*-methyl and *p*-methoxycalix[4]arenes,[54] but these data must be viewed with skepticism in view of the subsequent investigation of these products.[5,49]

[52] C. D. Gutsche and D. Stewart, unpublished observation.
[53] H. Erdtman, F. Haglid, and R. Ryhage, *Acta Chem. Scand.*, **1964**, *18*, 1249.
[54] T. B. Patrick and P. A. Egan, *J. Org. Chem.*, **1977**, *42*, 382; *ibid.*, **1977**, *42*, 4280.

Kämmerer and coworkers[43,45,47,56] have reported mass spectral data for many of the *p*-alkylcalixarenes synthesized by the Hayes and Hunter stepwise method, including the calix[7]arene which showed a parent ion *m/e* at 883. They observed[56] that the cyclic oligomers preferentially lose methyl or *tert*-butyl groups and conserve their ring structure in the mass spectral ion beam, whereas the corresponding linear oligomers preferentially cleave into their phenolic units. They have noted, however[56] that a calix[5]arene shows its strongest signal at *m/e* 480, apparently resulting from the extrusion of one of its aryl moieties to form the cyclic tetramer. This is in accord with the postulate that the cyclic tetramer is the thermodynamically most stable of the cyclic oligomers. Even the cyclic tetramer can undergo fragmentation in the ion beam, and signals corresponding to the extrusion of one, two, and three aryl moieties are observed. Bridging at the 'upper rim' of the calixarene appears to confer even greater stability on the calix, for Böhmer's arricho-arenes (see Figure 2.9) give spectra that are dominated by the parent ion to a greater extent than the corresponding acyclic counterparts or the analogous unsubstituted calixarenes.[57]

Though generally not as revealing of structural detail as *X*-ray crystallography or NMR spectroscopy, mass spectrometry nevertheless can often play a critical role. The host–guest features of the carcerands synthesized by Cram and coworkers,[58] for example, provide a case in point, as discussed in Chapter 6; another example is the structure of a resorcinol-derived calixarene containing twelve ferrocenyl moieties (see Chapter 7) which was confirmed by Beer and Keefe[59] by the observation of a signal at *m/e* 2921 for the parent ion.

3.5   TLC and HPLC Analysis of Calixarenes

Thin layer chromatography (TLC), introduced about mid-century, along with the related technique of high pressure (or high performance) liquid chromatography (HPLC) which is a more recent arrival, play an important part in modern organic chemistry. They have been of particular value in the calixarene field in the analysis of various reaction mixtures. In comparing the behavior of *p*-isopropylphenol, *p-tert*-butylphenol, *p-tert*-pentylphenol, and *p*-(1,1,3,3-tetramethylbutyl)phenol in the one-step procedures (*i.e.* Modified Zinke–Cornforth, Modified Petrolite, Standard Petrolite), Gutsche and coworkers[60] made extensive use of an HPLC method worked out by the Petrolite research group.[61] It employs a silica gel column and an eluant con-

[55] G. Happel, B. Mathiasch, and H. Kämmerer, *Makromol. Chem.*, **1975**, *176*, 3317.
[56] H. Kämmerer and G. Happel in '*Weyerhauser Science Symposium on Phenolic Resins*, 2', Tacoma, Washington, 1979, Weyerhaeuser Publishing Co., Tacoma, **1981**, p. 143.
[57] V. Böhmer, H. Goldmann, and W. Vogt, *J. Chem. Soc., Chem. Commun.*, **1985**, 667.
[58] D. J. Cram, S. Karbach, Y. H. Kim, L. Baczynskyj, K. Marti, R. M. Sampson, and G. W. Kalleymeyn, *J. Am. Chem. Soc.*, **1988**, *110*, 2554.
[59] P. D. Beer and A. D. Keefe, *J. Inclusion Phenom.*, **1987**, *5*, 499.
[60] B. Dhawan, S.-I. Chen, and C. D. Gutsche, *Makromol. Chem.*, **1987**, *188*, 921.
[61] F. J. Ludwig and A. G. Bailie, Jr., *Anal. Chem.*, **1984**, *56*, 2081.

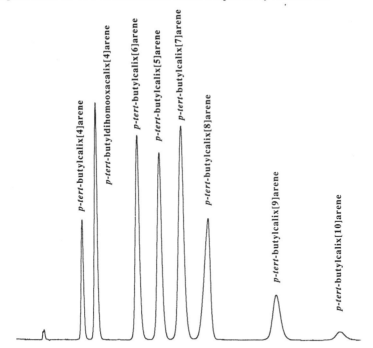

Figure 3.10 *Reversed-phase HPLC plot of a mixture of* p-tert-*butylcalixarenes*[52]

sisting of a mixture of 60% chloroform and 40% heptane, giving the elution sequence: cyclic octamer > cyclic heptamer > cyclic hexamer > cyclic tetramer > cyclic pentamer > dihomooxa compound. The Petrolite group has also devised an analytical procedure for linear as well as cyclic oligomers[62] that employs a reversed-phase column. In this case the order of elution with acetonitrile-methyl *tert*-butyl ether[52] as eluant, is: cyclic tetramer > di-homooxa compound > cyclic hexamer > cyclic pentamer > cyclic heptamer > cyclic ocatmer > cyclic nonamer > cyclic decamer, as shown in Figure 3.10. With both the standard and reversed phase columns the cyclic pentamer appears to be 'out of place', a phenomenon that may be correlated with its conformational and hydrogen bonding characteristics as discussed in the next chapter. High pressure liquid chromatography provides a particularly sensitive method for assessing the purity of the calixarene prepared by the one-step synthesis described in Chapter 2. Typical preparations showed the *p-tert*-butylcalix[4]arene to be *ca.* 99.5% pure (contaminated with *ca.* 0.5% of cyclic hexamer), the *p-tert*-butylcalix[6]arene *ca.* 99% pure (contaminated with *ca.* 0.5% of cyclic heptamer and 0.5% of an unidentified component), and the *p-tert*-butylcalix[8]arene *ca.* 100% pure (level of detection ≤ 0.1%).

[62] F. J. Ludwig and A. G. Bailie, Jr., *Anal. Chem.*, **1986**, *58*, 2069.

3.6   Concluding Remarks

One of the principal missions of chemistry is to separate mixtures of chemical species and to purify and characterize the individual components. Insight into chemical phenomena is gained far more easily with pure, well-defined compounds than with ill-defined mixtures. The phenol-derived calixarenes started life as mixtures, yielding only slowly and with many twists and turns to complete and correct analysis. In this respect it is interesting to compare calixarene chemistry with cyclodextrin chemistry where a rather similar situation existed. Cyclodextrins, which nature produces as mixtures of cyclooligomers, remained laboratory curiosities for half a century until procedures were developed in the 1950's for their separation, purification, and conclusive identification. Once they became readily available they proved to be attractive substances for study, and the literature of these compounds is now so extensive that it constitutes a major sub-discipline of chemistry, to be discussed in another volume in this series of monographs on supramolecular chemistry. Calixarenes have not yet reached this state of development, but as the following chapters will indicate, the easy accessibility of pure cyclooligomers has attracted a number of new researchers to the field and is taking calixarene chemistry along a path that in many respects resembles that followed by the cyclodextrins some years earlier.

CHAPTER 4

Shaping the Baskets: Conformations of Calixarenes

'A round man cannot be expected to fit into a square hole right away. He must have time to change his shape'

Mark Twain, *More Tramps Abroad*

Symmetry holds an incredible fascination for most people and has even been a dominant theme and guiding principle in cultures such as that of ancient Greece. Certainly, some of the attraction of supramolecular chemistry derives from the symmetry of many of its structures, as witness the beautiful cycles that characterize crown ethers and cyclodextrins. Calixarenes share this beauty, and the name calixarene was chosen because of the vase-like shape of the cyclic tetramer. It turns out, though, that the cyclic tetramer is amoeboid in character and can exist not only as a vase but in several other shapes as well. Calixarenes containing a greater number of aromatic rings are even more protean in character and can assume a still wider variety of forms. In this chapter we first look at the parent calixarenes (*i.e.* those molecules containing free intraannular or extraannular OH groups) and consider the ease with which transformations occur among the available shapes. Then attention is turned to calixarenes in which these transformations are constrained in one way or another, and the conformational consequences of these constraints are considered. The propensity of the calixarenes to assume various forms and the ability of the chemist to capture and freeze the system into one or another of these constitutes a particularly fascinating aspect of calixarene chemistry. Shaping the basket plays a potentially vital role in the design of calixarenes as enzyme mimics, for host–guest interactions depend on complementarity in shape as well as functionality.

4.1 Conformations of Flexible Calixarenes in the Solid State

This section deals with the *X*-ray crystallographic structures that have been established for free hydroxyl-containing calixarenes, compounds that generally show conformational mobility in solution and are frozen into a particular conformation only upon crystallization. The *X*-ray crystallographic structures of calixarenes that lack conformational mobility even in solution are discussed in subsequent sections.

4.1.1 Phenol-derived Calixarenes

X-Ray crystallography affords the best method for ascertaining conformations in the solid state, and this topic has already been dealt with to some extent in the previous chapter. As shown in the early investigations by the Parma group,[1,2] *p-tert*-butylcalix[4]arene and *p*-(1,1,3,3-tetramethylbutyl)calix[4]arene in the crystalline state both exist in the vase-like conformation (see Figure 3.1) that provides the basis for the generic name of the entire class. It is one of the four shapes that had already been realized as possibilities by Cornforth[3] whose ideas expanded on the earlier suggestions of Megson[4] and Ott and Zinke.[5] This conformer has been designated by Gutsche and coworkers[6] as the 'cone' form to distinguish it from the 'partial cone', '1,2-alternate', and '1,3-alternate' forms, as shown in Figure 4.1.

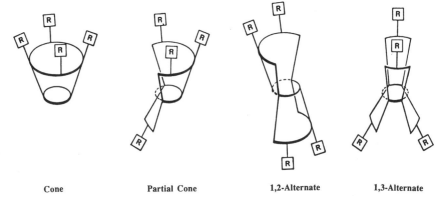

| Cone | Partial Cone | 1,2-Alternate | 1,3-Alternate |

Figure 4.1 *Pseudo 3-dimensional representations of the conformers of calix[4]arenes*

X-Ray crystallography of calix[5]arene[7] shows that it also exists in a cone conformation in the solid state (see Figure 3.1), although its shape begins to approach that of a shallow saucer. The calix[4]arenes and calix[5]arenes are identical with respect to the numbers of 'up–down' conformations that are possible, and for convenience the same descriptive names are used for both.

[1] G. D. Andreetti, R. Ungaro, and A. Pochini, *J. Chem. Soc., Chem. Commun.*, **1979**, 1005.
[2] G. D. Andreetti, A. Pochini, and R. Ungaro, *J. Chem. Soc., Perkin Trans. 2*, **1983**, 1773.
[3] J. W. Cornforth, P. D'Arcy Hart, G. A. Nicholls, R. J. W. Rees, and J. A. Stock, *Br. J. Pharmacol.*, **1955**, *10*, 73.
[4] N. R. L. Megson, *Oesterr. Chem.-Ztg.*, **1953**, *54*, 317.
[5] R. Ott and A. Zinke, *Oesterr. Chem.-Ztg.*, **1954**, *55*, 156.
[6] C. D. Gutsche, B. Dhawan, J. A. Levine, K. H. No, and L. J. Bauer, *Tetrahedron*, **1983**, *39*, 409.
[7] M. Coruzzi, G. D. Andreetti, V. Bocchi, A. Pochini, and R. Ungaro, *J. Chem. Soc., Perkin Trans. 2*, **1982**, 1133.

Calix[6]arenes can assume eight 'up–down' conformations, and because of the increased flexibility of this system additional conformations are possible in which one or more aryl rings assume a position approximately in the average plane of the molecule (designated as an 'out' alignment). Andreetti and coworkers[8] have shown that the crystalline compound assumes a shape close to what Gutsche has called a 'winged conformation', as illustrated in Figure 3.1.

An *X*-ray crystallographic determination of a calix[7]arene is not yet available, but that of *p-tert*-butylcalix[8]arene,[9] shows that the compound exists in a 'pleated loop' form, as pictured in Figure 3.1. In this structure the eight OH groups lie in a circular array which is an undulating 'pleated loop' that is ideally constituted for circular hydrogen bonding. For a calix[8]arene there are sixteen 'up–down' forms, as well as many others in which one or more aryl group assumes the 'out' alignment.

The calix[4]arenes and calix[5]arenes in the cone conformation nicely accommodate a circular array in which all of the OH groups are in the same plane, *i.e.* a homoplanar arrangement. The calix[8]arenes maximize intramolecular hydrogen bonding by assuming a conformation in which the OH groups lie above and below a plane, *i.e.* a heteroplanar arrangement. The calix[6]arenes fall between these two extremes; neither the homoplanar (*i.e.* a cone conformer) nor heteroplanar (*i.e.* a flattened 1,3,5-alternate conformer) arrangement of OH groups allows the most effective hydrogen bonding in this case.

4.1.2 Resorcinol-derived Calixarenes

The *X*-ray crystallographic structures of the resorcinol-derived calixarenes (see Section 3.1) establish their configurations as well as their conformations. As members of the calix[4]arene class, these products have the same four conformational possibilities that are pictured in Figure 4.1. The absence of intraannular OH groups, however, confers greater flexibility on the system and allows the aryl groups to assume 'out' alignments in addition to the four 'up–down' alignments. The names that have been chosen by the chemists working with these conformers are 'crown', 'boat', 'saddle', and 'chair'. However, to emphasize the similarity between the calixarenes carrying intraannular and extraannular OH groups, to avoid the use of the terms 'boat' and 'chair' which have precise applications in simple ring systems, and to promote a unified nomenclature we suggest that the conformers of the resorcinol-derived calixarenes be designated by the same names that are used for the phenol-derived calixarenes. Thus, 'crown' ≡ 'cone', 'chair' ≡ 'flattened partial cone', 'saddle' ≡ '1,3-alternate', and 'boat' ≡ 'flattened cone', as illustrated in Figure 4.2. The four *X*-ray structures that have been reported for members of this class are esters of calix[4]resorcinarene, so it may be debatable whether

[8] G. D. Andreetti, G. Calestani, F. Ugozzoli, A. Arduini, E. Ghidini, A. Pochini, and R. Ungaro, *J. Inclusion Phenom.*, **1987**, *5*, 123.
[9] C. D. Gutsche, A. E. Gutsche, and A. I. Karaulov, *J. Inclusion Phenom.*, **1985**, *3*, 447.

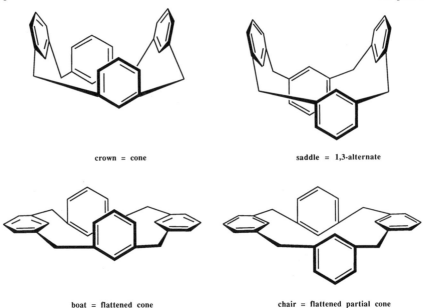

crown = cone   saddle = 1,3-alternate

boat = flattened cone   chair = flattened partial cone

Figure 4.2   *Conformations of the resorcinol-derived calixarenes*

they should be included in this section. However, as discussed in a later section, the esters do possess some conformational mobility in solution. The *X*-ray data for the crystalline samples show a flattened cone conformation for the octabutyrate[10] and octaacetate[11] of one of the isomers of C-*p*-bromo-phenylcalix[4]resorcinarene, as illustrated in Figure 3.3, and a flattened partial cone for another isomer. To what extent these conformations reflect those of the parent calixarenes is not certain, because intramolecular hydrogen bonding is absent in these octaesters.

4.2   Conformations of Flexible Calixarenes in Solution

4.2.1   Phenol-derived Calixarenes

All of the calixarenes containing free intraannular OH groups, including the calix[4]arenes, are conformationally mobile in solution at room temperature. Not surprisingly, the degree of mobility varies from one ring system to another, although not necessarily in the intuitively obvious fashion.

4.2.1.1   Conformations of the Calix[4]arenes. All four of the conformations for a calix[4]arene shown in Figure 4.1 are accessible by rotations of the aryl

[10] B. Nilsson, *Acta Chem. Scand.*, **1968**, *22*, 732.
[11] K. J. Palmer, R. Y. Wong, L. Jurd, and K. Stevens, *Acta Crystallogr.*, **1976**, *B32*, 847.

groups around the axis that passes through the *meta* carbon atoms bonded to the bridging methylene groups. The magnitude of the barrier to this rotation was overestimated by Cornforth and others, leading to the postulate of stable conformers. That the rotation is actually rather fast on the laboratory time-scale was first demonstrated by Kämmerer and coworkers[12,13] in their early studies employing temperature-dependent ^{1}H NMR measurements, and Munch's concurrent work[14] (see Chapter 1) appeared to provide further support to this concept. However, Munch's experiments involved a compound that at the time was thought to be a calix[4]arene but was later shown to be a calix[8]arene.[2,15] Extensions of these early studies have been carried out more recently by Gutsche and Bauer.[16,17]

The resonance arising from the methylene protons of a calix[4]arene appears as a pair of doublets at temperatures below *ca.* 20 °C and as a sharp singlet at temperatures above *ca.* 60 °C if the spectrum is measured in a nonpolar solvent such as chloroform. This behavior is interpreted in terms of a cone conformation that interconverts slowly on the NMR time-scale at the lower temperature but rapidly at the higher temperature. At the coalescence temperature of *ca.* 47 °C (at 100 MHz) the rate of interconversion can be calculated to be *ca.* 150 s^{-1}. Electronically neutral *para*-substituents have been shown to have a small though perceptible effect on the barrier to interconversion.[17] Thus, in chloroform solution the free energy of activation for *p-tert*-butylcalix[4]arene is 15.7 kcal/mole and that for *p-tert*-pentylcalix[4]arene is 14.5 kcal/mole. The several other calix[4]arenes that were investigated, including *p*-isopropyl-, *p*-(1,1,3,3-tetramethylbutyl)-, *p*-allyl-, and *p*-phenyl-, fall between these values, as shown by the data in Table 4.1. Measurements at 300 MHz for selected examples in this table gave $\Delta G^{\ddagger}$ values very close to those calculated from the coalescence data at 100 MHz, although a slightly lower value of 13.2 kcal/mole was obtained for *p-tert*-butylcalix[4]arene in pyridine.

Solvent polarity generally has a greater influence on the rate of conformational inversion of calix[4]arenes than does the *para*-substituent. In nonpolar solvents such as chloroform, toluene, benzene, bromobenzene, and carbon disulfide the barriers to rotation range from *ca.* 13.7 kcal/mole (calix[4]arene in benzene) to 15.7 kcal/mole (*p-tert*-butylcalix[4]arene in chloroform). This difference has been attributed[17] to the differing abilities of the calixarenes to form *endo*-calix complexes (see Chapter 6) with the solvent, *X*-ray crystallographic studies showing that the *tert*-butyl compound forms a much stronger solid-state complex than a calixarene carrying a hydrogen in the *para*-position.[1,2] In more polar solvents such as acetone and acetonitrile a significant decrease in the inversion barrier is typically observed, and this

[12] H. Kämmerer, G. Happel, and F. Caesar, *Makromol. Chem.*, **1972**, *162*, 179.
[13] G. Happel, B. Mathiasch, and H. Kämmerer, *Makromol. Chem.*, **1975**, *176*, 3317.
[14] J. H. Munch, *Makromol. Chem.*, **1977**, *178*, 69.
[15] C. D. Gutsche, R. Muthukrishnan, and K. H. No, *Tetrahedron Lett.*, **1979**, 2213.
[16] C. D. Gutsche and L. J. Bauer, *Tetrahedron Lett.*, **1981**, *22*, 4763.
[17] C. D. Gutsche and L. J. Bauer, *J. Am. Chem. Soc.*, **1985**, *107*, 6052.

Table 4.1 *Coalescence temperatures at 100 MHz and free energies of activation for conformational inversion of calix[4]arenes[17]*

para-Substituent	Deuterated solvent	T_c/°C	$\Delta G^{\ddagger}$/kcal mole⁻¹	para-Substituent	Deuterated solvent	T_c/°C	$\Delta G^{\ddagger}$/kcal mole⁻¹
tert-Butyl	dimethylformamide	60	15.7	Allyl	methylene chloride	42	15.0
	chloroform	52			chloroform	37	
	acetone	50			tetrahydrofuran	15	
	bromobenzene	43	15.2		pyridine	7.5	
	toluene	39	14.9		dimethylformamide	5	
	carbon disulfide	36	14.9		acetone	5	13.5
	benzene	35	14.8		acetonitrile	2	13.3
	tetrahydrofuran	15		tert-Octyl	chloroform	30	14.6
	pyridine	15	13.7		bromobenzene	24	14.3
H	methylene chloride	37.5			toluene	28	14.4
	chloroform	36	14.9		pyridine	−13	12.4
	bromobenzene	23	14.1	tert-Pentyl	chloroform	27	14.5
	toluene	18	13.9		bromobenzene	25	14.3
	benzene	15	13.8		toluene	36	14.8
	acetonitrile	0	13.3	Isopropyl	chloroform	33	14.8
	acetone	−5	13.1		bromobenzene	32	14.6
	pyridine	−22	11.8		toluene	30	14.4
Phenyl	chloroform	44	15.3	Benzoyl	chloroform	33	14.9
	bromobenzene	36	14.9	Hydroxyethyl	chloroform	44	15.3
	acetone	8	13.8	Cl, tert-Butyl	acetone	18	13.8
	pyridine	−2	12.8	Bromo	chloroform	38	15.0
					pyridine	2	13.0

becomes particularly pronounced in the basic, hydrogen-bonding solvent pyridine. For example, the inversion barrier for *p-tert*-butylcalix[4]arene falls from 15.7 kcal/mole in chloroform to 13.4 kcal/mole in pyridine; that of calix[4]arene falls from 14.7 kcal/mole in chloroform to 11.8 kcal/mole in pyridine. It is postulated[16,17] that the solvent effect, especially in the case of pyridine, is the result of disruption of the intramolecular hydrogen bonding that is a contributing force in maintaining the calixarene in the cone conformation.

The amino group constitutes an exception to the general rule that *para*-substituents have relatively little effect on the barrier to conformational inversion. As discussed in Section 3.3.4, calixarenes are stronger acids than their monomeric phenol counterparts, strong enough in fact to transfer a proton to an amine. Thus, in certain solvents calixarenes carrying amine functions in the *para*-positions exist as zwitterions.[18] Since calixarene oxyanions are conformationally less flexible than the fully protonated species (see Section 4.4.2), the zwitterions show higher rotational barriers than would be predicted solely on the basis of the bulk of the *p*-amino moiety. For the zwitterion to be stable, however, the solvent must have appreciable polarity. Thus, as shown in Table 4.2, coalescence temperatures are *lower* for *p*-dimethylaminocalix[4]arene in chloroform and bromobenzene than in acetonitrile, DMF, and DMSO — just the opposite of what is observed for the *para*-substituted calixarenes listed in Table 4.1.

It is interesting to speculate on the pathway for conformational inversion in the calix[4]arenes. One view[12] is that it may occur by way of a 1,3-alternate conformer which, being symmetrical, can revert to the cone and inverted cone conformers with equal facility. This route, which has been called the

Table 4.2 *Coalescence temperatures at 100 MHz and free energies of activation for the conformational inversion of calix[4]arenes carrying amino substituents in the* para-*position*[18]

Substituent	Deuterated solvent	$T_c/°C$	$\Delta G^{\ddagger}/\text{kcal mole}^{-1}$
Me_2NCH_2-	chloroform	52	15.7
	bromobenzene	46	15.4
	acetonitrile	67	16.2
	pyridine	40	14.7
	DMF	71	6.3
	CF_3CO_2H	27	
	DMSO	82	16.9
$C_6H_5CH_2N\diagup\diagdown NCH_2-$	chloroform	57	16.0
	DMSO	85	17.1
$H_2NCH_2CH_2-$	DMSO	90	17.3
$N\underset{CH}{\overset{CH=CH}{<}}NCH_2-$	DMSO	62	15.9

[18] C. D. Gutsche and K. C. Nam, *J. Am. Chem. Soc.*, **1988**, *110*, 6153.

'broken chain pathway',[17] requires the complete disruption of the hydrogen bonding of the circular array of the initial cone conformation, although some hydrogen bonding is regained in the 1,3-alternate conformer. As an alternative, a 'continuous chain pathway' has been suggested[17] in which the aryl groups swing through the annulus in tandem, leading to an activated complex that resembles a skewed 1,2-alternate conformer, as illustrated in Figure 4.3. Although hydrogen bond stretching is necessary to achieve this structure, it appears to represent a lower energy pathway that might provide a reasonable explanation for the similarity of conformational barriers between the calix[4]arenes and calix[8]arenes, as discussed in the next section.

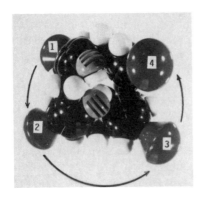   

"Continuous Chain" Pathway   "Broken Chain" Pathway

Figure 4.3   *Space-filling models of the intermediate states for the 'broken chain' and 'continuous chain' pathways for the conformational interconversion of calix[4]arenes*

Along with the conformations of the calixarene macrocycle itself, additional conformational possibilities can result from the presence of certain functional groups. For example, *p*-benzoylcalix[4]arene permits various orientations of the benzoyl groups, providing conformers which Gutsche and Pagoria[19] have named 'open calix', 'closed calix', 'spiral calix', and 'stacked' as illustrated in Figure 4.4.

4.2.1.2 Conformations of Calix[8]arenes. Rather than discussing conformations in the sequence of the ring size, the calix[8]arenes are considered in

[19] C. D. Gutsche and P. F. Pagoria, *J. Org. Chem.*, **1985**, *50*, 5795.

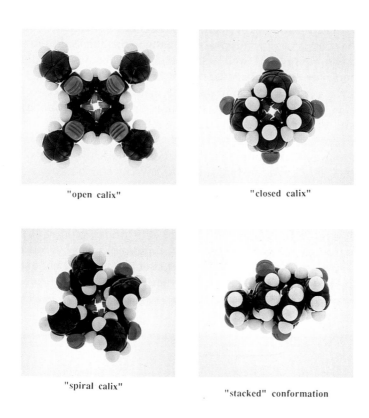

"open calix"   "closed calix"

"spiral calix"
                           "stacked" conformation

Figure 4.4   *CPK models of the cone conformations of* p-benzoylcalix[4]arene *in the 'open calix', 'closed calix', 'spiral calix', and 'stacked' conformations*

tandem with the calix[4]arenes because of the remarkable similarity of the inversion barriers in the two series. As previously noted, sixteen 'up–down' conformations of the calix[8]arenes can exist, along with numerous others in which one or more aryl groups assume an 'out' alignment. Space-filling models suggest that the cyclic octamer should be much more flexible than the cyclic tetramer, and one might expect this to be reflected in its ^{1}H NMR characteristics. Yet, in nonpolar solvents, calix[8]arenes possess temperature-dependent ^{1}H NMR spectra virtually identical with those of calix[4]arenes.

As illustrated in Figure 4.5, the coalescence temperatures are almost identical, and the $\Delta G^{\ddagger}$ values differ by a small fraction of a kcal/mole. This was initially interpreted in terms of a 'pinched' conformation[16] in which the cyclic octamer puckers to create a pair of circular arrays of hydrogen bonds with four OH groups in each array, a conformation that is invoked in the 'molecular mitosis' process discussed in Chapter 2 (see Figure 2.16). However, since *p-tert*-butylcalix[8]arene assumes a pleated loop conformation in the solid state, it is reasonable to assume that this is the dominant conformation in solution as well. The interconversion between pleated loop conformers, the $\Delta G^{\ddagger}$ values for which are shown in Table 4.3, probably occurs by a 'continuous chain' pathway in which intramolecular hydrogen bonding is maintained to the maximum extent. The similarity in the magnitude of the interconversion barriers for the calix[4]arenes and calix[8]arenes lends credence to this pathway for the smaller ring system. But, it is a curious quirk of nature that the barriers in these two systems, seemingly so different in inherent flexibility, should have inversion barriers so close to one another. It is a coincidence that impeded the correct assignment of structure of the calix[8]arenes for several years.

The similarity in the temperature-dependent ^{1}H NMR spectra of the cyclic tetramers and cyclic octamers disappears in pyridine solution. It is assumed that in both cases the pyridine disrupts the intramolecular hydrogen bonding, leading to a weakening of the forces that maintain the cone and pleated loop structures, respectively. In the cyclic tetramers there are other forces in addi-

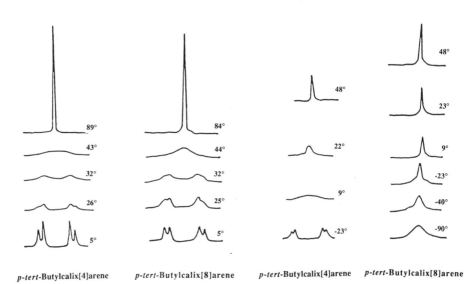

p-tert-Butylcalix[4]arene	*p-tert*-Butylcalix[8]arene	*p-tert*-Butylcalix[4]arene	*p-tert*-Butylcalix[8]arene
CDCl$_3$ solution		pyridine-d$_5$ solution	

Figure 4.5 *Temperature-dependent ^{1}H NMR spectra of* p-tert-*butylcalix[4]arene and* p-tert-*butylcalix[8]arene in CDCl$_3$ and pyridine-d$_5$*[16]

Table 4.3 *Coalescence temperatures (at* 100 *MHz) and free energies of activation for the conformational inversion of calix[8]arenes*

Substituent	Deuterated solvent	$T_c/°C$	$\Delta G^{\ddagger}/\text{kcal mole}^{-1}$
tert-Butyl	chloroform	53	15.7
	bromobenzene	43	15.1
	toluene	36	14.7
	carbon disulfide	32	14.6
	benzene	33	14.6
	acetone	−48	10.6
	pyridine	below −90	<9
H	pyridine	below −90	<9
Phenyl	pyridine	below −90	<9
tert-Octyl	chloroform	43	15.2
	bromobenzene	53	15.6
	toluene	23	14.1
	pyridine		<9
tert-Pentyl	chloroform	59	16
	bromobenzene	53	15.6
	toluene	36	14.8
Isopropyl	chloroform	38	15.0
	bromobenzene	31	14.6
	toluene	20	13.9

tion to hydrogen bonding that favor the cone conformation, and these are weak or absent in the cyclic octamers.

4.2.1.3 Conformations of the Calix[6]arenes. As mentioned above, calix[6]arenes can exist in eight 'up–down' conformations as well as others in which one or more of the aryl groups assume an 'out' alignment. Temperature-dependent ^{1}H NMR measurements show that in nonpolar solvents the cyclic hexamers are more flexible than either the cyclic tetramers or cyclic octamers, as indicated by a coalescence temperature of 11 °C (at 100 MHz) for *p-tert*-butylcalix[6]arene in chloroform solution ($\Delta G^{\ddagger}$ of 13.3 kcal/mole). The low-temperature spectrum of a calix[6]arene, pictured in Figure 4.6, displays a twelve line pattern arising from the methylene protons that can be interpreted as three sets of overlapping quartets. This is commensurate with conformations in which the molecule transannularly pinches in a fashion that (a) places two of the aryl groups in 'out' alignments and the other four in nonequivalent 'up' and/or 'down' positions (the twisted 'winged' conformation) or (b) places three contiguous aryl groups 'up' and the other three 'down' (the 'hinged' conformation). Space-filling molecular models show that in both of these conformations the six OH groups can form two clusters in which circular hydrogen bonding might be possible, although the $H \cdot\cdot O \cdot\cdot H$ angle is very acute. It is also more or less commensurate with the conformation known to exist in the solid state (see Figure 3.1).

para-Substituents have relatively little influence on the barrier to conformational inversion of the calix[6]arenes, as shown by the data in Table 4.4.

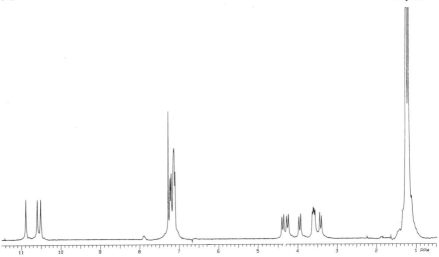

Figure 4.6   *¹H NMR spectrum* (300 MHz) *of* p-tert-*butylcalix[6]arene in CDCl₃ at* − 50 °C

Table 4.4   *Coalescence temperatures (at* 100 MHz*) and free energies of activation for the conformational inversion of calix[6]arenes*

Substituent	Deuterated solvent	$T_c/°C$	$\Delta G^{\ddagger}/\text{kcal mole}^{-1}$
tert-Butyl	chloroform	11	13.3
	bromobenzene	− 12	13.0
	acetone	− 40	11.1
	pyridine	− 54	9.0
H	chloroform	3	13.2
	acetone	− 49	9.5
	pyridine	− 70	< 9
tert-Octyl	chloroform	12	13.6
Allyl	chloroform	− 2	13.0

The magnitude of the 'pyridine effect' for a calix[6]arene falls between that of a calix[4]arene and a calix[8]arene.

4.2.1.4   Conformations of Calix[5]arenes, Calix[7]arenes, and Calix[9]arenes. Until recently, the relative inaccessibility of the odd-numbered calixarenes has relegated them to a back seat in the study of conformational behavior. *p*-*tert*-Butylcalix[5]arene, prepared by the one-step procedure, shows a coalescence temperature in chloroform solution of − 2 °C at 100 MHz, corresponding to an activation barrier of 13.2 kcal/mole. Kämmerer and coworkers[20] reported similar data for a calix[5]arene containing *tert*-butyl

²⁰ H. Kämmerer, G. Happel, and B. Mathiasch, *Makromol. Chem.*, **1981**, *182*, 1685.

and methyl groups, synthesized by a stepwise procedure. The ^{1}H NMR pattern arising from the methylene groups is commensurate with a cone conformation, which *X*-ray crystallography has shown to be its shape in the solid state.

A temperature-dependent ^{1}H NMR spectrum of a calix[7]arene carrying six *p*-methyl groups and one *p-tert*-butyl group (see Figure 2.5), synthesized by a stepwise method,[21] is reported to have a coalescence temperature of − 10 °C at 100 MHz, corresponding to an inversion barrier of 12.3 kcal/mole. A calix[7]arene carrying seven *p-tert*-butyl groups, obtained from a one-step reaction, shows a coalescence temperature of 15 °C at 300 MHz, corresponding to the somewhat higher inversion barrier of 13.4 kcal/mole. *p-tert*-Butylcalix[9]arene is almost identical in its behavior with the cyclic heptamer, showing a coalescence temperature of 17 °C at 300 MHz, corresponding to a $\Delta G^{\ddagger}$ of 13.5 kcal/mole. Thus, the cyclic pentamer, cyclic hexamer, cyclic heptamer, and cyclic nonamer possess almost identical barriers to conformational inversion, placing the cyclic tetramer and cyclic octamer in a separate class of particularly stable conformers. Almost certainly this is due to the especially favorable intramolecular hydrogen bonding that exists in these two calixarenes, in contrast to the other calixarenes where the inability to assume a tightly hydrogen-bonded cone conformation or the inability to assume a continuous zig-zag conformation reduces the effectiveness of intramolecular hydrogen bonding. Space-filling models of the cyclic heptamer, for example, suggest that six of the aryl residues can form a continuous pleated ribbon but that considerable strain is introduced if the seventh aryl residue is brought into the loop. It has been predicted[17] that when *X*-ray crystallographic data become available for this compound they will show the structure to be substantially planar but with one of the aryl groups somewhat out of the plane.

4.2.1.5 Conformations of Oxacalixarenes. The larger annulus of dihomooxacalix[4]arenes as compared with calix[4]arenes leads to greater conformational flexibility, lower coalescence temperatures, and lower rotational barriers. A chloroform solution of *p-tert*-butyldihomooxacalix[4]arene (see Figure 2.24) gives rise to a room temperature ^{1}H NMR spectrum containing a broad singlet arising from the methylene protons. At lower temperatures this resolves into an eight line pattern that can be interpreted as two sets of pairs of doublets, commensurate with a cone conformation. The coalescence temperatures for these two sets of resonances are slightly different, *viz.* − 8 and − 2 °C, corresponding to $\Delta G^{\ddagger}$ values of 12.9 and 13.0 kcal/mole, respectively, which is *ca.* 2.9 kcal/mole lower than in *p-tert*-butylcalix[4]arene. Additional changes take place in the methylene resonances as the temperature is lowered further, the downfield set of doublets changing to a broad singlet at − 60 °C and then at − 90 °C resolving into a pair of doublets at approximately the same position as those observed in the higher temperature spectrum. This behavior has been explained[17] as a flexing of the

[21] H. Kämmerer and G. Happel, *Makromol. Chem.*, **1980**, *181*, 2049.

CH_2OCH_2 bridge to give conformers in which the oxygen is 'inside' and 'outside' the cavity.

The temperature-dependent ^1H NMR spectrum of the dihomooxacalix[4]arene in pyridine gives a coalescence temperature of $-32\,°C$, corresponding to an inversion barrier of 10.0 kcal/mole and placing it between the calix[4]arenes and the calix[6]arenes. The 3 kcal/mole lower value in chloroform is only slightly greater than that for *p-tert*-butylcalix[4]arene, however, and suggests that the role played by intramolecular hydrogen bonding is approximately the same in the two systems.

When a pair of CH_2OCH_2 bridges is present in a calix[4]arene ring (see Figure 2.26) the conformational flexibility is further increased, and *p-tert*-butyltetrahomodioxacalix[4]arene shows a coalescence temperature in chloroform at $-24\,°C$ corresponding to an inversion barrier of 11.9 kcal/mole. In pyridine the coalescence temperature falls below $-70\,°C$, again illustrating the importance of intramolecular hydrogen bonding in maintaining conformational integrity. The presence of three CH_2OCH_2 bridges in a calixarene leads to still greater flexibility, the ^1H NMR spectrum of *p-tert*-butylhexahomotrioxacalix[3]arene (see Figure 2.26) in chloroform showing only a singlet for the methylene resonance down to $-90\,°C$.

4.2.1.6   Conformational Comparisons of Phenol-derived Calixarenes. The conformational flexibility of calixarenes and oxacalixarenes carrying intraannular OH groups depends on the size of the macrocyclic ring which, in turn, influences the nature of the intramolecular hydrogen bonding. The relation between OH stretching frequencies in the infrared spectra, OH resonance positions in the ^1H NMR spectra, and the inversion barrier for various calixarenes is shown in Table 4.5. The calix[4]arenes, calix[5]arenes, and dihomooxacalix[4]arenes maximize their intramolecular hydrogen bonding by assuming the cone conformation, while tetrahomodioxa-

Table 4.5   *Stretching frequencies (IR), chemical shifts (^1H NMR), and free energies of activation (in $CDCl_3$) for conformational inversion of* p-tert-*butylcalixarenes and oxacalixarenes*

Compound	ν_{OH}/cm^{-1}	δ_{OH}	$\Delta G^{\ddagger}/kcal\ mole^{-1}$
calix[4]arene	*3138 **3164	10.2	15.7
calix[5]arene	*3290 **3303	8.0	13.2
calix[6]arene	*3152 **3127	10.5	13.3
calix[7]arene	*3149 **3180	10.3	13.4
calix[8]arene	*3190 **3258	9.6	15.7
calix[9]arene	**3274	9.8	13.5
dihomooxacalix[4]arene	3300	9.0, 9.7	12.9
tetrahomodioxacalix[4]arene	3700	9.0	11.9
hexahomotrioxacalix[3]arene	3410	8.5	<9

*in CCl_4 solution (S. W. Keller, G. M. Schuster, and F. L. Tobiason, *Polym. Mater. Sci. Eng.*, **1987**, *57*, 906)
**in KBr pellet (D. Stewart and C. D. Gutsche)

calix[4]arenes do so by assuming a flattened cone conformation. Calix[8]arenes adopt a pleated loop conformation, and calix[7]arenes probably adopt conformations approaching this shape. Calix[6]arenes are thought to exist either as a 'hinged' or 'winged' conformation in solution. Thus, as the size of the annulus of the calixarene increases, the preferred conformations become increasingly planar. Conformational inversion occurs with varying degrees of ease, the rotation barriers in a nonpolar solvent decreasing in the order: calix[4]arenes = calix[8]arenes > calix[5]arenes = dihomooxacalix[4]-arenes = calix[6]arenes = calix[7]arenes = calix[9]arenes > tetrahomodioxacalix[4]arenes > hexahomotrioxacalix[3]arenes. In pyridine solution the intramolecular hydrogen bonding is disrupted, and the inversion barrier becomes primarily a function of ring size, the rotational barriers decreasing in the order: calix[4]arene > calix[5]arene > dihomooxacalix[4]arene > calix[6]arene > tetrahomodioxacalix[4]arene > calix[8]arene > hexahomotrioxacalix[3]arene. Except for amino groups, the nature of the *para*-substituent plays only a minor role in determining the magnitude of the rotational barriers. In Figure 4.7 the conformations of five of the calixarenes are pictured in space-filling molecular models.

4.2.2 Resorcinol-derived Calixarenes

Although the parent compounds obtained from the resorcinol–aldehyde condensations have conformational flexibility, the possibility for cone–cone interconversion comparable to that in the phenol-derived calixarenes exists only in certain cases. The groups attached to the bridging methylenes (introduced by the aldehyde used in the condensation) bring configurational as well as conformational considerations into play. With respect to configuration there are four possible combinations when the system is in the cone conformation, as shown in Figure 4.8. These are designated as **A** (*cis, cis, cis, cis*), **B** (*trans, cis, cis, trans*), **C** (*trans, trans, trans, trans*), and **D** (*trans, cis, trans, cis*). If each of these conformers is inverted, **A** (a,a,a,a) transforms to the diastereoisomer **E** (e,e,e,e), and **B** (e,a,a,a) transforms to the diastereoisomer **F** (a,e,e,e). However, **C** and **D** lead to structures identical with the originals, so there is a total of six diastereoisomeric forms. The diastereoisomeric possibilities for the other conformers can be enumerated in similar fashion, giving a total count of more than two dozen.

All of the studies of the conformational mobility of these species in solution have been carried out on esters of the parent calixarenes. In a carefully detailed analysis of the temperature-dependent ^{1}H NMR spectra of the octabutyrates of C-*p*-bromophenylcalix[4]resorcinarene, Högberg assigned conformations as well as configurations.[22] For the diastereoisomer that is the product of thermodynamic control and which had been shown by X-ray crystallography to possess the all-*cis* configuration he observed a pair of

[22] A. G. S. Högberg, *J. Am. Chem. Soc.*, **1980**, *102*, 6046.

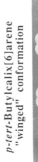

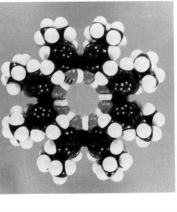

p-tert-Butylcalix[6]arene
"winged" conformation

p-tert-Butylcalix[8]arene

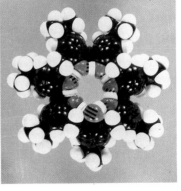

p-tert-Butylcalix[5]arene

p-tert-Butylcalix[7]arene

p-tert-Butylcalix[4]arene

p-tert-Butylcalix[6]arene
"pleated loop" conformation

Figure 4.7 *Space-filling molecular models of phenol-derived calixarenes showing the arrangement of intramolecularly hydrogen bonded hydroxyl groups*

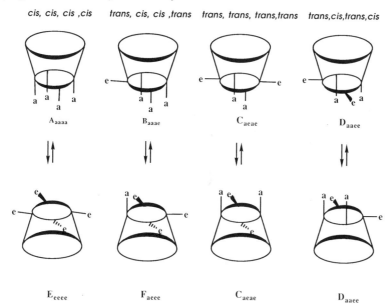

Figure 4.8   *Diastereoisomers of a resorcinol-derived calixarene in the cone conformation. The lower case letters a and e designate the alignment of the bonds
carrying the R groups (C-2, C-8, C-14, C-20) as axial or equatorial*

coalescence temperatures that correspond to $\Delta G^{\ddagger}$ values of 18.96 and 12.9
kcal/mole. The higher value is assigned to a pseudorotation that involves
interconversion between the degenerate forms of the all-axial flattened cone
conformation (A_{aaaa} and A'_{aaaa}), as illustrated in Figure 4.9. The barrier to this
pseudorotation arises from the crowding together of the four axial R groups
as the system passes through the full cone conformation, which probably
represents the transition state for the process. The lower value is assigned to
a rotation of the aryl groups (Ar) around the single bonds that join them to
the methylene bridge carbons. Not surprisingly, C-phenylcalix[4]resorcin-
arene gives the same value for the high-energy barrier, while C-methylcalix[4]resorcinarene gives the somewhat lower values of 14.4 and 15.2
kcal/mole for the octaacetate and octapropionate esters, respectively.[23]

To the product of kinetic control from *p*-bromobenzaldehyde and resorcinol, Högberg assigned a *trans,cis,trans,cis* configuration (**D**) and a flattened
partial cone conformation in which the R groups at C-2, C-8, C-14, and
C-20 all assume axial alignments. In contrast to the flattened cone conformer
from **A**, the flattened partial cone conformer from **D** has a *non-degenerate*
pseudorotation, *i.e.* the interchange of the positions of the four calixarene
aryl groups in D_{aaaa} produces D'_{aeae} in which the alignments of the R groups at
C-2 and C-14 have changed from axial to equatorial. The similiarity in the
values of the low-energy barriers in the kinetic and thermodynamic products

[23] A. G. S. Högberg, *J. Org. Chem.*, **1980**, *45*, 4498.

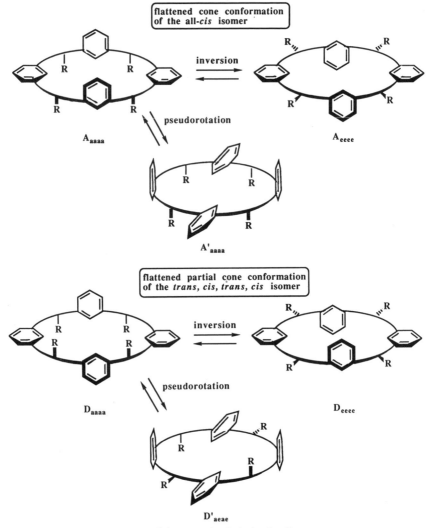

Figure 4.9   *Conformations of the resorcinol-derived calixarenes*

led Högberg to postulate that both of the isomers with which he was dealing
have all of the R groups in axial alignments, a postulate that is commensurate
with the *trans,cis,trans,cis* stereochemistry for the kinetic product[10] that was
subsequently confirmed by X-ray crystallography. A third resorcinol-derived
calixarene isomer has recently been isolated as the octacetate from condensations with heptaldehyde and dodecylaldehyde.[24] On the basis of ^{1}H NMR
data, including NOE and T_1 measurements, it is postulated to have a
trans,cis,cis,trans configuration and a 1,2-alternate conformation. The
dominant influence in all three of these diastereoisomers appears to be the

[24] L. Abis, E. Dalcanale, A. Du vosel, and S. Spera, *J. Org. Chem.*, **1988**, *53*, 5475.

preference for the R groups at C-2, C-8, C-14, and C-20 to assume the axial alignment.

4.2.3 Conformations of non-Hydroxylic Calixarenes

The conformational behavior of the calixarene derived from mesitylene and formaldehyde (see Figure 2.3)[25] has been studied.[26] The methyl groups are enough larger than hydroxyl groups to prevent the rotation of the aryl moieties around the *meta–meta* axes, and as a consequence the compound shows no temperature-dependent ^{1}H NMR spectral behavior but is frozen in the 1,3-alternate conformation. Thus, it perhaps belongs more properly in the next section of this chapter. On the other hand, a non-hydroxylic calixarene that does show conformational mobility is the bis-tetrahydropyrimidone calixarene **85** in Figure 2.29.[27] At − 40 °C this compound displays two sets of pairs of doublets in the ^{1}H NMR spectrum that arise from the CH_2 protons and correspond to a major and a minor conformer. The signals from the two conformers coalesce at 0 °C to a single pair of doublets and at 70 °C to a singlet. The authors speculate that the major isomer is the cone conformer and the minor isomer is the 1,3-alternate conformer.

4.3 Conformationally Immobile Calixarenes

Donald Cram joined the staff of the University of California at Los Angeles at mid-century, and for over two decades he concentrated his research efforts, with great ingenuity and success, on various aspects of physical organic chemistry. Then, in 1970 his emphasis shifted to the crown ethers, which had been announced only a short time earlier by Charles Pedersen[28] of the duPont Company. Cram and his group proceeded to explore crown ether chemistry, again with characteristic ingenuity and attention to detail, hoping to create compounds capable of engaging in what Cram calls 'host–guest' chemistry. So successful was this endeavor that in 1987 he shared the Nobel Prize with Charles Pedersen and Jean-Marie Lehn, whose work at the University of Strasbourg has set new and exciting goals for what Lehn calls 'receptor–substrate' chemistry or, more generally, 'supramolecular chemistry'. Cram has defined a conformationally immobile basket or 'cavitand' as a *synthetic* compound that contains an 'enforced cavity'[29] large enough to engulf ions or molecules.[30,31] Although the calix[4]arenes meet Cram's criterion of synthetic accessibility, their conformational mobility makes them ephemeral rather than permanent baskets. To convert them into the latter, to make them constant calixes, it is necessary to freeze them either

[25] J. L. Ballard, W. B. Kay, and E. L. Kropa, *J. Paint Technol.*, **1966**, *38*, 251.
[26] S. Pappalardo, F. Bottino, and G. Ronsisvalle, *Phosphorus Sulfur*, **1984**, *19*, 327.
[27] J. A. E. Pratt, I. O. Sutherland, and R. F. Newton, *J. Chem. Soc., Perkin Trans. 1*, **1988**, 13.
[28] C. J. Pedersen, *J. Am. Chem. Soc.*, **1967**, *89*, 2495, 7017; idem, ibid., **1970**, *92*, 386, 391.
[29] R. C. Helgeson, J.-P. Mazaleyrat, and D. J. Cram, *J. Am. Chem. Soc.*, **1981**, *103*, 3929.
[30] J. R. Moran, S. Karbach, and D. J. Cram, *J. Am. Chem. Soc.*, **1982**, *104*, 5826.
[31] D. J. Cram, *Science*, **1983**, *219*, 1177.

Donald Cram

in a cone or a partial cone conformation. This has been accomplished in a variety of ways, one of the most elegant being that devised by Cram for the resorcinol-derived calixarenes as discussed below in Section 4.3.2. `

4.3.1   Conformational Freezing *via* 'Lower Rim' Functionalization

4.3.1.1   Esters and Ethers. Since the pathway for conformational inversion in the calix[4]arenes involves rotation of the aryl groups in a direction that brings the OH groups through the annulus of the macrocyclic ring, the most obvious way to curtail this motion is to replace the OH group with larger moieties. Conversion into the ester or ether is the easiest way to accomplish this, and dozens of such compounds have now been prepared and characterized. Considering the conformational possibilities of the calix[4]arenes, it is apparent that the esters and ethers can exist in any one of the four conformations: cone, partial cone, 1,2-alternate, or 1,3-alternate, as pictured in Figure 4.1. *X*-Ray crystallography provides the surest way to discern which of these conformers is present in a particular case, and there are several examples of its application to calix[4]arene derivatives. The earliest one involves the tetraacetate of *p-tert*-butylcalix[4]arene (**1a**) which the Parma group[32] showed to be fixed in the partial cone conformation, as illustrated in Figure 3.2. The carbonate ester (**1b**), on the other hand, was shown by

[32] C. Rizzoli, G. D. Andreetti, R. Ungaro, and A. Pochini, *J. Mol. Struct.*, **1982**, *82*, 133.

1a   R = OCOCH$_3$

1b   R = OCO$_2$Et

1c   R = OCH$_2$CO$_2$Et

1d   R = OCH$_2$CO$_2$Bu^t

1e   R = OCH$_2$CONEt$_2$

1f   R = OCH$_2$COCH$_3$

1g   R = SCONMe$_2$

McKervey and coworkers[33] to assume a cone conformation as do most of the ethers for which *X*-ray structures have been obtained. For example, compounds **1c**, **1d**, **1e**, and **1f** have all been shown to fall in this category.[34–36] The only example of an unbridged calix[4]arene that assumes the 1,2-alternate conformation is the tetra-N-N-dimethylthioureido compound **1g** synthesized by the St Louis group.[37]

X-Ray crystallography demands not only the appropriate experimental facilities but a suitable crystal, and frequently one or the other of these requisites is missing. Fortunately, there is a good alternative for establishing the conformations of calix[4]arenes, for each of the conformers displays a dis-

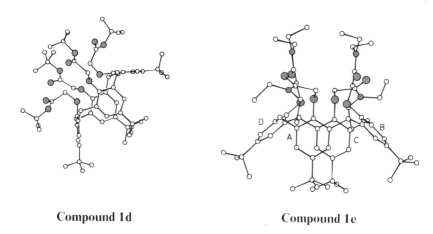

Compound 1d   Compound 1e

Figure 4.10   *X-Ray crystallographic structures of calix[4]arene derivatives*

[33] M. A. McKervey, E. M. Seward, G. Ferguson, and B. L. Ruhl, *J. Org. Chem.*, **1986**, *51*, 3581.
[34] R. Ungaro, A. Pochini, and G. D. Andreetti, *J. Inclusion Phenom.*, **1984**, *2*, 199.
[35] G. Calestani, F. Ugozzoli, A. Arduini, E. Ghidini, and R. Ungaro, *J. Chem. Soc., Chem. Commun.*, **1987**, 344.
[36] M. A. McKervey, E. M. Seward, G. Ferguson, B. Ruhl, and S. J. Harris, *J. Chem. Soc., Chem. Commun.*, **1985**, 388.
[37] C. D. Gutsche, J. S. Rogers, G. F. Stanley, and P. K. Sujeeth, unpublished work.

Table 4.6   *^{1}H NMR patterns for the CH$_2$ protons of calix[4]arenes in various conformations*

Conformation	^{1}H NMR Pattern
Cone	One pair of doublets
Partial cone	Two pairs of doublets (ratio 1:1) or one pair of doublets and one singlet (ratio 1:1)
1,2-Alternate	One singlet and two doublets (ratio 1:1)
1,3-Alternate	One singlet

tinctive pattern in the methylene region of the ^{1}H NMR, as shown in tabular form in Table 4.6 and by the spectra in Figure 4.11.

Ungaro and coworkers[38] and Gutsche and coworkers[6] have employed this simple, useful, and highly dependable probe to study the conformations of the calix[4]arene ethers and esters shown in Figure 4.12. In the light of the formation of flattened 1,3-alternate conformations from some of the resorcinol-derived calixarenes, it was initially anticipated that the 1,3-alternate conformation would be favored in the phenol-derived calix[4]arene derivatives. However, it turns out that these calixarenes are more often captured in the cone or partial cone conformation. Thus, methylation and ethylation of **2** and **3** and allylation of **2** yield the corresponding ethers in which the ethyl and allyl ethers are partial cones. Surprisingly, the methyl ethers **5a** and **6a** remain almost as conformationally mobile as the parent compounds, providing another example of how space-filling models tend to overemphasize conformational barriers. The recent report[39] that the azulocalix[4]arene tetramethyl ether (see Figure 2.11) is frozen in the 1,3-alternate conformation presents an interesting contrast to these data.

Conversion of **3** into the tetraallyl ether (**6c**), tetrabenzyl ether (**6d**), or tetratrimethylsilyl ether (**6e**) freezes the calixarene in the cone conformation. Similarly, the tetrabenzyl ether (**5d**), tetratosylate (**5f**), tetratrimethylsilyl ether (**7e**), and tetratosylate (**7f**) are generated in the cone conformation. Fortunately, derivatization reactions frequently produce good yields of a single conformer, although mixtures are sometimes encountered. For example, acetylation of compound **38** in Figure 2.12 yields a 1,3-alternate and two partial cone conformers.[40,41] However, conversion of this same calixarene into the tetratrimethylsilyl derivative yields only the cone conformer. Still other examples of calix[4]arenes in the cone conformation are compounds **1b**, **1c**, **1d**, and **1e** for which *X*-ray structures have been obtained.

The conformations of several non-bridged, partially O-substituted calix[4]arenes have been studied. The trimethyl ether of *p-tert*-butyl-

[38] V. Bocchi, D. Foina, A. Pochini, and R. Ungaro, *Tetrahedron*, **1982**, *38*, 373.
[39] T. Asao, S. Ito, and N. Morita, *Tetrahedron Lett.*, **1988**, *29*, 2889.
[40] K. H. No and C. D. Gutsche, *J. Org. Chem.*, **1982**, *47*, 2713.
[41] The formation of two partial cone conformers is the result of the lower symmetry of this calixarene, making possible the existence of six conformers, *viz.* one cone, two partial cones, two 1,2-alternates, and one 1,3-alternate.

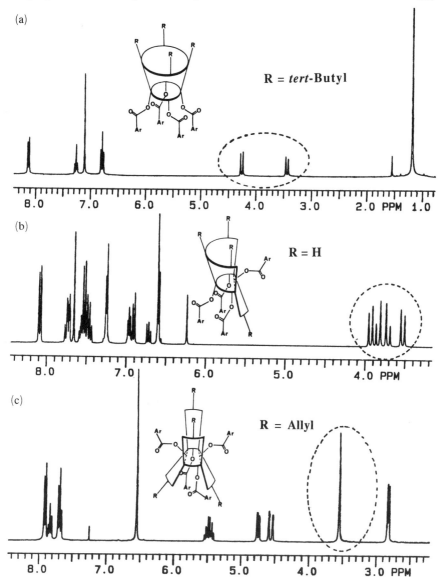

Figure 4.11 *^{1}H NMR spectra at* 300 MHz *of calixarene tetrabenzoates of (a) p-tert-butylcalix[4]arene in the cone conformation; (b) calix[4]arene in the partial cone conformation; (c) p-allylcalix[4]arene in the 1,3-alternate conformation*

calix[4]arene is somewhat less conformationally flexible than the tetramethyl ether,[42] showing well resolved ^{1}H NMR resonances in the δ3—4 region with a multiplicity and a chemical shift (see below) commensurate with a flattened

[42] C. D. Gustche, B. Dhawan, J. A. Levine, K. H. No, and L. J. Bauer, *Tetrahedron*, **1983**, *39*, 409.

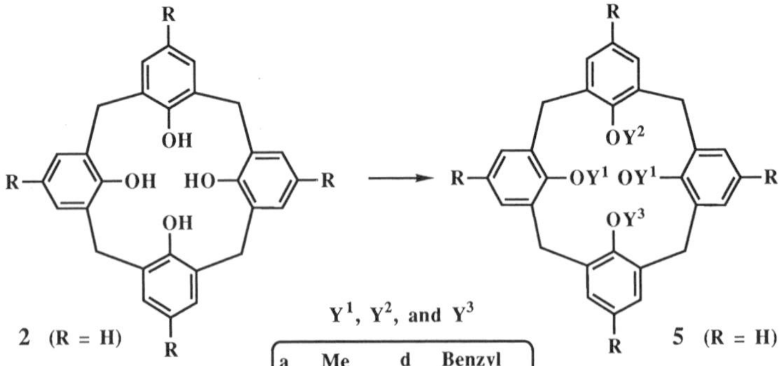

Y^1, Y^2, and Y^3			
a	Me	d	Benzyl
b	Et	e	SiMe$_3$
c	Allyl	f	SO$_2$C$_7$H$_7$

2 (R = H)

3 (R = *tert*-Bu)

4 R = Allyl)

5 (R = H)

6 (R = *tert*-Bu)

7 (R =Allyl)

Figure 4.12   *Calix[4]arene esters and ethers*

partial cone conformation. Not surprisingly, the tribenzoate of calix[4]arene (**6**, Y^1 = Y^2 = COC$_6$H$_5$, Y^3 = H) is also conformationally inflexible. That it is frozen in a partial cone conformation is indicated by its ^{1}H NMR spectrum as well as an *X*-ray crystallographic determination of the corresponding allyl ether[43] (**6**, Y^1 = Y^2 = COC$_6$H$_5$, Y^3 = allyl). The 1,3-dimethyl ether and the 1,3-dibenzyl ether of *p-tert*-butylcalix[4]arene[42] (**6**, Y^1 = H, Y^2 = Y^3 = Me and CH$_2$C$_6$H$_5$, respectively) are even less conformationally flexible, showing ^{1}H NMR spectra commensurate with either a distorted cone or a flattened 1,3-alternate conformation. The OH groups in both of these compounds exchange their protons only slowly, suggesting that they are protected by being buried in the cavities of the molecules. Probably they are intramolecularly hydrogen bonded, a feature that may explain why the partially methylated compounds are so much less flexible than their fully methylated analogs. In similar fashion, the diaroyl esters of calix[4]arenes are conformationally stable, existing in a flattened cone or a flattened 1,3-alternate conformation depending on the *para*-substituents on the 'upper rim'.[44] Even the mono-allyl ether of calix[4]arene (**5**, Y^1 = Y^2 = H, Y^3 = allyl),[45] is more conformationally rigid than might have been expected, retaining a sharp pattern of ^{1}H NMR resonances from the methylene protons at 60 °C.

The magnitude of the chemical shift $(\Delta\delta)$ between the high- and low-field pairs of resonances arising form the methylene protons of a calix[4]arene is dependent on the conformation. As illustrated in Figure 4.13, which shows the methylene group of a calix[4]arene oriented so that the hydrogens project

---

[43] A. Vrielink, P. W. Codding, C. D. Gutsche, and L.-g. Lin, *J. Inclusion Phenom.*, **1986**, *4*, 199.
[44] C. D. Gutsche and K. A. See, to be published.
[45] C. D. Gutsche and L.-g. Lin, *Tetrahedron*, **1986**, *42*, 1633.

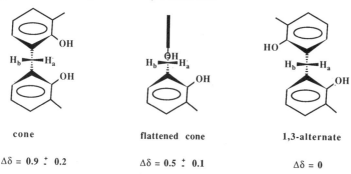

cone	flattened cone	1,3-alternate

$\Delta\delta = 0.9 \pm 0.2$ $\Delta\delta = 0.5 \pm 0.1$ $\Delta\delta = 0$

Figure 4.13   *Conformational dependence of the chemical shift $(\Delta\delta)$ between the high- and low-field pairs of resonances of the methylene protons in the calix[4]arenes*

toward the viewer and the aryl rings project away from the viewer, $\Delta\delta$ is generally *ca.* 0.9 ± 0.2 for a system in the cone conformation and zero for a system in the 1,3-alternate conformation if the measurement is made in a nonpolar solvent such as $CDCl_3$. In between these extremes, in the 'flattened' conformations, $\Delta\delta$ assumes intermediate values of *ca.* 0.5 ± 0.1, thus providing an indication of the degree to which the system departs from true cone or 1,3-alternate conformation. The $\Delta\delta$ values are also dependent on solvent, however, and are generally larger in polar solvents such as pyridine and DMSO than in nonpolar solvents. For example, *p-tert*-butylcalix[4]arene has a $\Delta\delta$ of 0.73 in $CDCl_3$ and 1.25 in pyridine-d_5.

It is a fortunate circumstance that the calix[4]arenes can be fixed in cone and partial cone conformations, for this establishes them as 'enforced cavities' and enhances their attractiveness for the synthesis of enzyme mimics. The conformation in which a particular calix[4]arene is captured upon esterification or etherification is predictable more on the basis of analogy than on insight, although the experiments that are described in the next chapter provide at least partial answers to the intriguing question of how reagents, conditions, and the *para*-substituent of the calix[4]arene all play a part in determining the conformational outcome of a derivatization reaction.

Calix[6]arenes possess larger annuli than calix[4]arenes and, as a consequence, form esters and ethers that have greater conformational mobility and that require larger moieties to curtail mobility. For example, the methyl ether of *p-tert*-butylcalix[6]arene is conformationally mobile even at $-60\,°C$; the hexaacetate has a coalescence temperature of $25\,°C$ (corresponding to $\Delta G^{\ddagger}$ of 14 kcal/mole); and the hexatrimethylsilyl ether has a coalescence temperature of $61\,°C$ (corresponding to $\Delta G^{\ddagger}$ of 16.1 kcal/mole). The ${}^1H$ NMR spectrum of the hexatrimethylsilyl ether in chloroform is commensurate with a structure in which the aryl groups are 'up–out–up–down–out–down', the same conformation that has been established by *X*-ray crystallographic determination on the hexa(2-methoxyethyl)ether of *p-tert*-butylcalix[6]arene[46] and the hexamethyl ether of *p*-allylcalix[6]arene[47] (see Figure 3.2).

The activation barriers for conformational inversion of the methyl ethers, benzyl ethers, and acetates of calix[6]arenes are essentially independent of the identity of the *para*-substituent. This is reasonable if one assumes that the inversion pathway involves motion of the aryl rings in the direction that brings their oxygen moieties through the annulus, as discussed above for the calix[4]arenes. It turns out, though, that in the case of the hexatrimethylsilyl ethers of calix[6]arenes the rate of conformational inversion *is* dependent on the *p*-substituent. This has been interpreted in terms of an opposite rotatory motion in which the aryl groups swing in the direction that brings their *para*-substituents through the annulus, as illustrated in Figure 4.14. Space-filling molecular models indicate that this is a plausible pathway if the moieties attached to the oxygens are sufficiently large, as is the case with the hexatrimethylsilyl ethers.

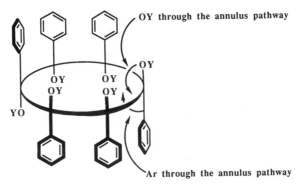

Figure 4.14   *Pathways for conformational interconversion of calix[6]arenes*

A few partially O-substituted calix[6]arenes have been prepared and their conformations studied.[48] Among these is the 1,2,4,5-tetra-(*p*-nitrobenzoate) of *p-tert*-butylcalix[6]arene which has the possibility of fourteen 'up–down' conformations. Fortunately, the simplicity of its ^{1}H NMR spectrum, shown in Figure 4.15, permits the solution state structure to be assigned either conformation **A** or **B**. On the basis of transient NOE measurements a slightly distorted, conformationally equilibrating structure **B** appears to be preferred in which two of the hydroxyl-bearing rings are tipped inward so that their *tert*-butyl groups occupy the 'cavity' on the two faces of the molecule.

The calix[8]arenes have even larger annuli and require correspondingly larger groups at *both* ends of the aromatic rings to curtail conformational interconversion. The octatrimethylsilyl ether of *p-tert*-butylcalix[8]arene, for

[46] R. Ungaro, A. Pochini, G. D. Andreetti, and P. Domiano, *J. Inclusion Phenom.*, **1985**, *3*, 35.
[47] L.-g. Lin, G. F. Stanley, and C. D. Gutsche, unpublished work.
[48] C. D. Gutsche and J. S. Rogers, unpublished work.

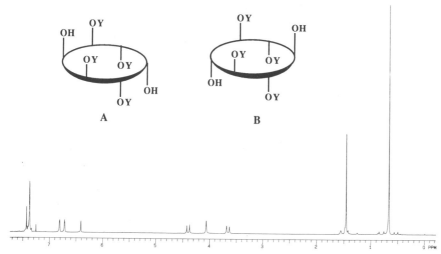

Figure 4.15 *¹H NMR spectrum and conformation of a calix[6]arene tetraester*[48]

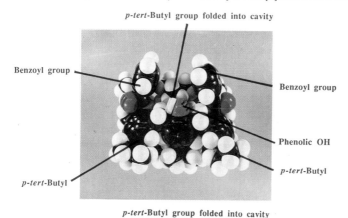

CPK Model of a 1,2,4,5-tetraester of a calix[6]arene

example, has a coalescence temperature of $-5\,°C$ (corresponding to an inversion barrier of 13.2 kcal/mole), showing that under ordinary conditions it is a conformationally mobile system. To date, there is no published work dealing with the freezing of the conformation of these large ring calixarenes by simple ester or ether formation, although, this has been accomplished by the use of bridging moieties, as discussed in the next section.

4.3.1.2 Conformational Freezing via Covalent Bridges. Bridging provides another device for conformational fixing, and it has been applied to the calix-arenes at the 'lower rim' as well as the 'upper rim'. The first example of 'lower rim bridging' was published in 1983 by Ungaro and coworkers who treated *p-tert*-butylcalix[4]arene with the ditosylate of pentamethylene glycol and

8a   R = (CH$_2$)$_5$

8b   R = (CH$_2$CH$_2$O)$_4$

obtained a calixarene bridged between transannular oxygens[49] (8a). They provided a second example a year later[50] in which a tetraethyleneoxy bridge was transannularly inserted to give the crown ether-like structure 8b. A more recent, and very interesting, example[51] makes use of a metallocene bridge, introduced by treatment of *p-tert*-butylcalix[4]arene with 1,1'-bis(chloro-carbonyl)metallocene (M = Fe and Ru) in the presence of triethylamine. The ^{1}H NMR spectra of these metallocenes (9, M = Fe and Ru) show a pair of CH$_2$ doublets that sharpen and remain resolved on heating, behavior that is commensurate with a flattened cone conformation. On cooling to − 100 °C, however, another pair of AB doublets appears along with four singlets from the metallocene protons, four singlets from the aromatic protons, three singlets from the *tert*-butyl groups, and two singlets from the OH groups. This pattern is interpreted by the authors in terms of either a partial cone con-formation or the freezing of the rotational motion of the carbonyls of the ester groups. It is our view that the latter hypothesis is more likely the correct one. A more elaborate illustration of 'lower rim' bridging was recently provided by compound 10 where the combined talents at the laboratories of Parma and Enschede were brought to bear on its synthesis from the dimethyl ether of *p-tert*-butylcalix[4]arene[52] and a hemispherand. Compound 10, designated as a calixspherand and verified by X-ray crystallography of a Na$^+$ complex, is dis-cussed in Chapter 6 for its interesting complexation characteristics. The ^{1}H NMR spectrum of the uncomplexed compound indicates that it exists in the partial cone conformation. Advantage has been taken by Gutsche and coworkers[53] of the formation of 1,3-diaroylates from calix[4]arenes for build-

[49] G. Alfieri, E. Dradi, A. Pochini, R. Ungaro, and G. D. Andreetti, *J. Chem. Soc., Chem. Commun.*, 1983, 1075.

[50] R. Ungaro, A. Pochini, and G. D. Andreetti, *J. Inclusion Phenom.*, 1984, 2, 199.

[51] P. D. Beer and A. D. Keefe, *J. Inclusion Phenom.*, 1987, 5, 499.

[52] D. N. Reinhoudt, P. J. Dijkstra, P. J. A. in't Veld, K. E. Bugge, S. Harkema, R. Ungaro, and E. Ghidini, *J. Am. Chem. Soc.*, 1987, 109, 4761.

[53] C. D. Gutsche, M. Iqbal, K. C. Nam, K. A. See, and I. Alam, *Pure Appl. Chem.*, 1988, 60, 483.

ing 'lower rim' bridges. In a sequence of reactions that is described in the next chapter, they prepared a 'double cavity' calixarene (**11**) in which two bridges are inserted between the amino groups of 3,5-diaminobenzoyl moieties. In contrast to the calixspherand **10**, which assumes the partial cone conformation, solutions of compounds **11a** and **11b** both possess ^{1}H NMR spectral patterns characteristic of a cone conformation. These 'lower rim'-bridged calixarenes are shown in Figure 4.16.

4.3.1.3 Conformational Freezing via Oxygen–Metal Bridges. The interaction of calixarenes with inorganic compounds gives entities that are often referred to as complexes. However, they are treated in this chapter as compounds containing oxygen–metal–oxygen bonds, because they constitute a form of 'lower rim' bridging.

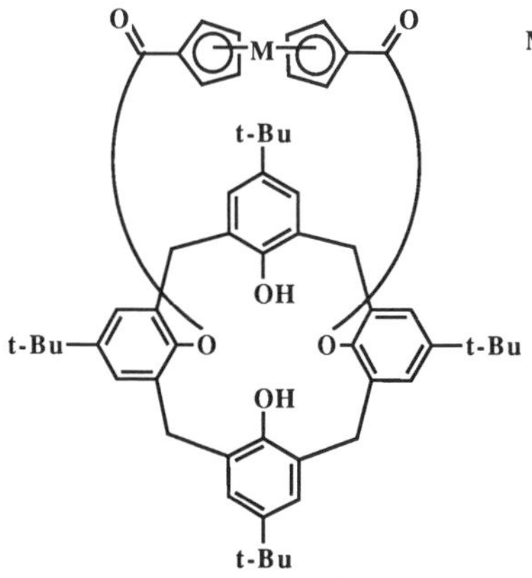

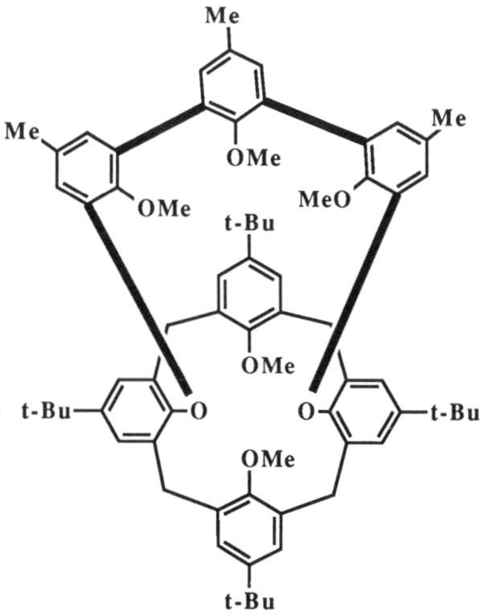

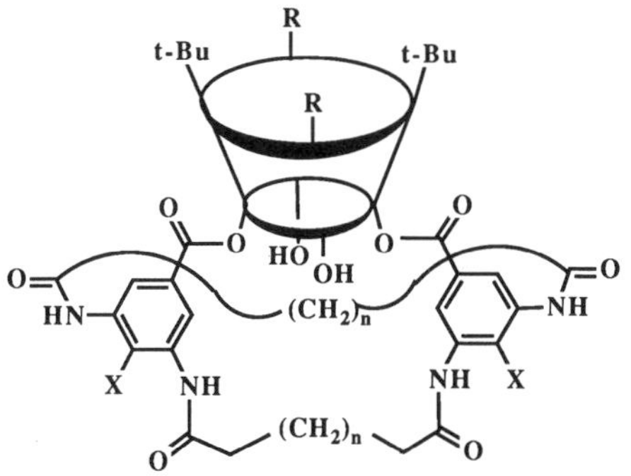

Figure 4.16 *'Lower rim' bridged calix[4]arenes*

Philip Power and his coworkers at the University of California at Davis[54] were the first to explore the interaction of calixarenes with transition metal ions in their search for unusual coordination and reactivity patterns. In one experiment in this study they treated *p-tert*-butylcalix[4]arene with $Ti(NMe_2)_4$ and obtained a titanium alkoxide in which two titanium atoms are sandwiched between a pair of cyclic tetramers, *viz.*

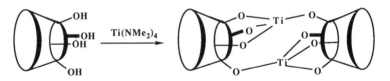

X-Ray crystallography established the structure shown in Figure 4.17 in which the two calixarene units assume a partial cone conformation to most effectively accommodate the tetrahedral coordination geometry around the titanium atoms. Crystallization of the titanium complex from toluene yields a compound containing three toluene molecules, one of which is in an *endo*-calix position. The structural rigidity of the titanium complex is indicated by its temperature-independent ^1H NMR spectrum in chloroform which shows a pair of doublets from the CH_2 protons that remain invariant from -30 to $+80\,^\circ C$.

In similar fashion *p-tert*-butylcalix[4]arene was treated with $Fe[N(SiMe_3)_2]_3$ and $Co[N(SiMe_3)_2]_2$ to give low but reproducible yields of $[Fe(NH_3)$ (calix-arene$-OSiMe_3)_2]$ and $[Co_3(calixarene-OSiMe_3)_2(THF)]$, respectively. *X*-Ray

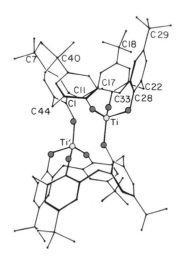

Figure 4.17 X-*Ray crystallographic structure of a titanium–calix[4]arene compound*[54]

[54] M. M. Olmstead, G. Sigel, H. Hope, X. Xu, and P. P. Power, *J. Am. Chem. Soc.*, **1985**, *107*, 8087.

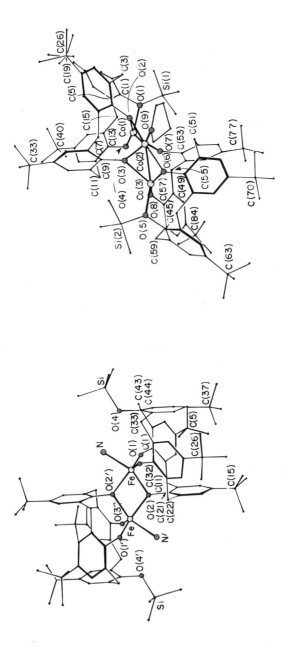

Figure 4.18 X-Ray crystallographic structure of Fe-calix[4]arene and Co-calix[4]arene compounds[54]

crystallography of these complexes produced the structures pictured in Figure 4.18 which show the configuration of the iron to be distorted trigonal pyramidal and that of the cobalt to even be less conventional, all three of the cobalts being described as irregular; Co(1) and its symmetry partner Co(3), for example, have all of their ligands different from one another. Also, the bridging configuration of the —OSiMe₃ group is unique. The cobalt compound crystallizes with five molecules of toluene, two of which are engaged in *endo*-calix complexation.

Andreetti and coworkers[8] treated *p-tert*-butylcalix[6]arene with Ti(OPr)₄ and obtained an orange, crystalline solid whose X-ray crystallographic structure showed it to consist of two calixarene units in a cone conformation coordinated with two titanium atoms connected to one another by a Ti—O—Ti bond, as illustrated in Figure 4.19. The titanium atoms are penta-coordinated in a slightly distorted trigonal bipyramid. The solid state structure is retained in solution, as indicated by the ¹H NMR spectrum which contains three doublets for the equatorial protons of the methylene groups and three doublets for the axial protons. Thus, interaction with the titanium ion induces a significant molecular reorganization to a more cone-like conformation.

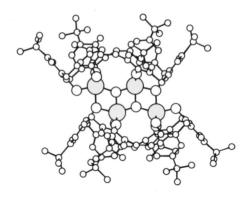

Figure 4.19   X-*Ray crystallographic structure of a titanium complex of* p-tert-*butylcalix[6]arene*[8]

Jack Harrowfield and his coworkers[55] in Australia have shown that *p-tert*-butylcalix[8]arene forms a series of homo- and hetero-bimetallic complexes with the lanthanide ions. A crystalline europium complex has been obtained[56] that is shown by X-ray crystallography to contain two europium atoms embedded in a 'pinched' conformer of the cyclic octamer. As the schematic and ORTEP representations in Figure 4.20 indicate, the complex contains one molecule of the cyclic octamer as a hexaanion (two OH groups

[55] B. M. Furphy, J. M. Harrowfield, and F. R. Wilner, to be published.
[56] B. M. Furphy, J. M. Harrowfield, D. L. Kepert, B. W. Skelton, A. H. White, and F. R. Wilner, *Inorg. Chem.*, **1987**, *26*, 4231.

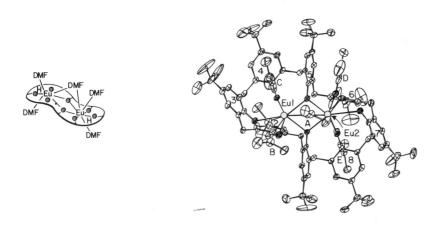

Figure 4.20 *Schematic and ORTEP representations of the di-europium complex of* p-tert-*butylcalix[8]arene*[56]

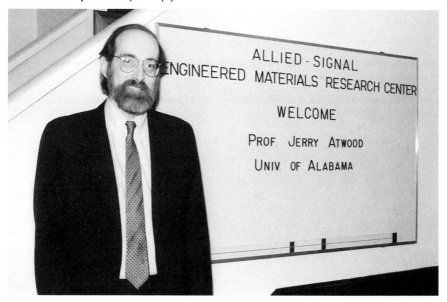

Jerry Atwood

remain), two europium atoms, and five DMF molecules (one in a bridging position and four singly coordinated to the europium atoms).

Jerry Atwood of the University of Alabama, an *X*-ray crystallographer with active interests in a variety of areas of inorganic chemistry, heard a lecture by Gutsche on calixarenes at a 1982 conference in Parma that piqued his interest. Subsequent application of his expertise to this area has brought new perspectives to the calixarene field and has produced some exotic examples of complex formation in the quest to discover ways in which catalytically active early transition metals can interact with macrocycles.

The first of the Atwood publications on calixarenes appeared in 1986 and describes the reaction of the hexamethyl ether of *p-tert*-butylcalix[6]arene with $TiCl_4$ in toluene solution.[57] The light red crystalline product is shown by *X*-ray crystallography to have the structure pictured in Figure 4.21 in which the calixarene has combined with a pair of bimetallic units ($Cl_3TiOTiCl_2$). This is interpreted in terms of (a) reaction of two of the OMe groups of the calixarene with two molecules of $TiCl_4$ to produce two calixarene—O—Ti bonds and (b) reaction of the resulting compound with two more molecules derived from $TiCl_4$ (along with, presumably, some adventitious water) to produce two Ti—O—Ti bonds. The conformation adopted by the calix[6]arene to provide the most effective coordination between the methoxyl groups and the Ti atoms is best described as a 'down, out, out, down, out, out' shape, which is different from that of the simple calix[6]arene ether whose *X*-ray crystallographic structure is shown in Figure 3.2.

In studying what they call 'the liquid clathrate phenomenon', Atwood and coworkers have explored the interaction of several calixarene methyl ethers with aluminum alkyls. A benzene–toluene solution of the tetramethyl ether of *p-tert*-butylcalix[4]arene, for example, interacts with Me_3Al to form large,

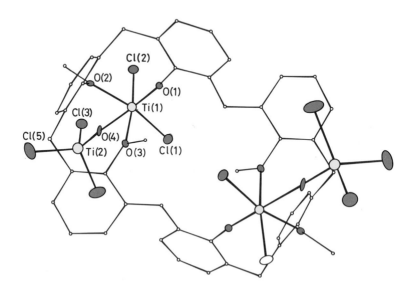

Figure 4.21   *The* p-tert-*butylcalix[6]arene hexamethyl ether–TiCl₄ 'complex'*[57]

[57] S. G. Bott, A. W. Coleman, and J. L. Atwood, *J. Chem. Soc., Chem. Commun.*, **1986**, 610.

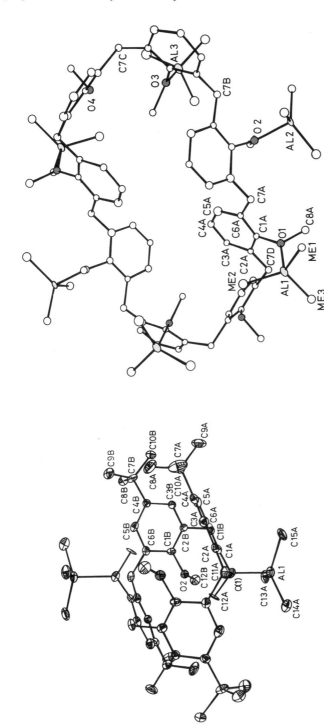

Trimethylaluminum complex of the octa-methyl ether of *p-tert*-butylcalix[8]arene

Trimethylaluminum complex of the tetra-methyl ether of *p-tert*-butylcalix[4]arene in the 1,2-alternate conformation

Figure 4.22 *Structures of the trimethylaluminum complexes of the methyl ethers of calixarenes*[58]

colorless crystals which are shown by *X*-ray crystallography[58] to have a structure in which a molecule of the calixarene in the rarely encountered 1,2-alternate conformation is combined with two molecules of Me_3Al, as pictured in Figure 4.22. Crystals similarly obtained from *p-tert*-butylcalix[4]arene tetramethyl ether but with $MeAlCl_2$ or $EtAlCl_2$, also are structures containing one molecule of calixarene and two molecules of the aluminum compound, but in both of these cases the calixarene assumes the more usual 1,3-alternate conformation. In a similar study with the octamethyl ethers of calix[8]arene and *p-tert*-butylcalix[8]arene,[59] the products are found to be structures containing six molecules of Me_3Al per calixarene molecule. The effect of the *p-tert*-butyl groups is to reduce the flexibility of the system. While the structure from the calix[6]arene has all six of the Me_3Al molecules on the outside, that from the *p-tert*-butylcalix[6]arene has two of them within the boundary of the macrocyclic ring. Also as part of the liquid clathrate study, Atwood *et al.* treated a toluene solution of the tetramethyl ether of *p-tert*-butylcalix[4]arene with Me_3Al along with sodium benzoate containing one equivalent of water.[60] The resulting material is a large, colorless, mildly air-sensitive crystal that was shown to consist of a complex anion containing Me_3Al and $C_6H_5CO_2^-$ and a complex cation containing the calixarene ether, Na^+, and a toluene molecule combined in the fashion illustrated in Figure 4.23. The sodium, which lies 0.44 Å out of the plane of the four oxygens of the methoxyl groups, strongly interacts with all of them, the average Na⋯O distance being 2.30 Å. The fifth coordination site is filled remotely by the complex anion.

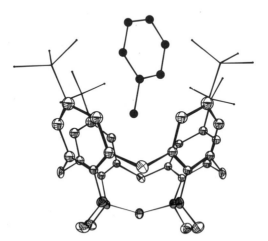

Figure 4.23   X-*Ray crystallographic structure of the* p-tert-*butylcalix[4]arene tetramethyl ether–toluene–Na+ compound*[60]

[58] S. W. Bott, A. W. Coleman, and J. L. Atwood, *J. Inclusion Phenom.*, **1987**, *5*, 747.
[59] A. W. Coleman, S. G. Bott, and J. L. Atwood, *J. Inclusion Phenom.*, **1987**, *5*, 581.
[60] S. G. Bott, A. W. Coleman, and J. L. Atwood, *J. Am. Chem. Soc.*, **1986**, *108*, 1709.

4.3.2 Conformational Freezing *via* 'Upper Rim' Functionalization

The 'upper rim' plays no direct part in the conformational inversion process in the calix[4]arenes, and *para*-substituents have only a slight effect on the rate (see Section 4.2.1.1). The 'upper rim' of a calix[6]arene assumes importance only if the substituents on the oxygen atoms are large enough to preclude the 'OY through the annulus pathway', as portrayed in Figure 4.14. In calixarenes of any size, however, 'upper rim' bridging can provide a device for inhibiting conformational inversion, although to date it has been explored only with the calix[4]arenes. Calixarenes bridged on the 'upper rim' have been synthesized by Böhmer and coworkers,[61-64] who added polymethylene spanners of various lengths between transannular *para*-positions of a calix[4]arene (**12**) (see Figure 4.24), as discussed in Chapter 2 (see Figure 2.9). The conformational freezing that results from the spanner is reflected in the ^{1}H NMR spectra of these compounds which show essentially no solvent or temperature dependence. Thus, spectra measured in chloroform, nitrobenzene, and pyridine at temperatures ranging from -63 to $+182\,°C$ all display pairs of doublets arising from the methylene protons, characteristic of the cone conformation. Some interesting features of calix[4]arenes bridged in this fashion are discussed in Chapter 6. Another example of a calixarene bridged on the 'upper rim' is the dihomooxacalix[4]arene **13** (see Figure 4.24), synthesized as shown in Figure 2.25, in which a pair of adjacent rings are joined by an ethylene spanner. This compound likewise displays an ^{1}H

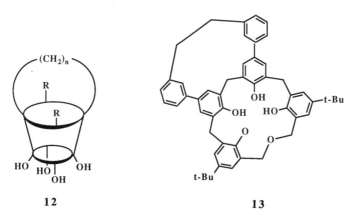

12   **13**

Figure 4.24   *'Upper rim' bridged calixarenes*

[61] V. Böhmer, H. Goldmann, and W. Vogt, *J. Chem. Soc., Chem. Commun.*, **1985**, 667.
[62] E. Paulus, V. Böhmer, H. Goldmann, and W. Vogt, *J. Chem. Soc., Perkin Trans. 2*, **1987**, 1609.
[63] H. Goldmann, W. Vogt, E. Paulus, and V. Böhmer, *J. Am. Chem. Soc.*, **1988**, *110*, 6811.
[64] V. Böhmer, H. Goldmann, R. Kaptein, and L. Zetta, *J. Chem. Soc., Chem. Commun.*, **1987**, 1358.

NMR spectrum in chloroform solution that shows a highly resolved pattern of resonances arising from the methylene protons (three non-equivalent sets produce a 12-line spectrum), indicative of severely curtailed conformational mobility.[65]

A particularly elegant example of 'upper rim' bridging has been realized by Cram and coworkers with the resorcinol-derived calixarenes. Treatment of the parent octahydroxy compound with bis-electrophiles (*e.g.* $ClCH_2Br$, $TsO(CH_2)_nOTS$, and Me_2SiCl_2) in polar aprotic solvents (*e.g.* THF or DMSO) containing a base (*e.g.* Et_3N or Cs_2CO_3) under high dilution conditions, introduces bridges between each of the four pairs of proximate OH groups to yield compounds **14** ($X = CH_2$, CH_2CH_2, $CH_2CH_2CH_2$, and Me_2Si)[29,66] (see Figure 4.25). In similar fashion[29] treatment with 2,3-dichloro-1,4-diazanaphthalene yields compound **15**, as illustrated in Figure 4.25. All of these bridged compounds are conformationally immobile with

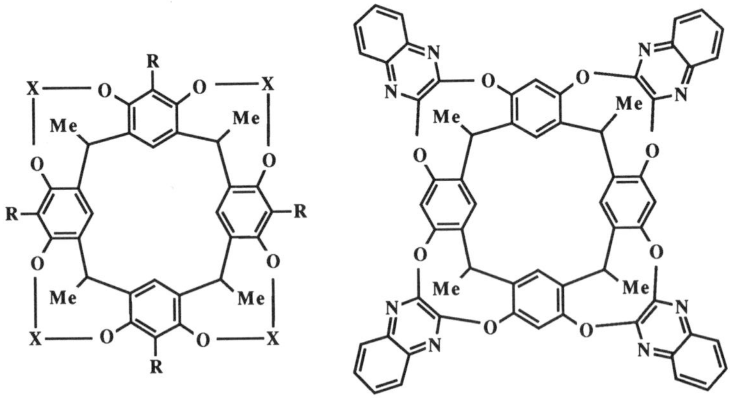

14   **15**

Figure 4.25   *'Upper rim' bridged resorcinol-derived calixarenes*[66]

respect to the calixarene ring system, although **15** shows a temperature-dependent ^{1}H NMR spectrum as a result of the flexing of the diazanaphthalene rings from an axial to an equatorial alignment. These compounds, which are examples of cavitands, show particularly fascinating complexation properties, as discussed in Chapter 6.

4.4   The Conformations of Calixarene Oxyanions

4.4.1   Calix[4]arene Oxyanions

The pK_a values for the calix[4]arenes, discussed in Chapter 3, have been shown by Shinkai and coworkers to be lower for pK_1 and higher for pK_4 than

[65] C. D. Gutsche, P. K. Sujeeth, and D. W. Bailey, unpublished experiments.
[66] D. J. Cram, S. Karbach, H.-E. Kim, C. B. Knobler, E. F. Maverick, J. L. Ericson, and R. C. Helgeson, *J. Am. Chem. Soc.*, **1988**, *110*, 2229.

CPK Model of a cavitand

might be expected by comparison with the values of the monomeric phenols. In a study of the effect of oxyanion formation on calixarene conformation Gutsche *et al.*[53] treated a number of calix[4]arenes with *n*-butyllithium in DMSO solution. The base was added in incremental amounts corresponding to one, two, three, and four equivalents per equivalent of calixarene, the presumption being that the mono-, di-, tri-, and tetra-anions of the calixarene would be generated. Ultraviolet spectral measurements showed a progressive increase in absorptivity at *ca.* 310 nm up to the addition of the fourth equivalent of base but no further change as more base was added. The ^{1}H NMR spectra of these solutions showed a well-resolved pair of doublets in the $\delta 2.5$—5.0 region for the one-equivalent and four-equivalent cases (characteristic of a cone conformation), a more complex pattern for the three-equivalent case (characteristic of a partial cone conformation), and a still more complex pattern for the two-equivalent case. The pattern of the two-equivalent solution is best interpreted as a superposition of the spectra of the one- and three-equivalent solutions, suggesting that it is a mixture of the mono- and tri-anions with little or no di-anion. The temperature dependencies of the ^{1}H NMR spectra reveal that the coalescence temperature for the mono-anion is only slightly higher than that of the parent calixarene, while that of the tri- and tetra-anions is considerably higher — in fact, above 140—160 °C. Complementary data were obtained by measuring the ^{7}Li NMR spectra which displayed a single resonance for the mono- and tetra-anions, three resonances for the trianion, and a pattern for the two-equivalent solution that is best interpreted as a superposition of the one- and three-equivalent solution. Lacking *X*-ray crystallographic information on the lithium oxyanions, one can only speculate about their structure. On the basis of measurements on CPK models the annulus at the 'lower rim' of the calix[4]arenes appears to be too small to complex Li$^+$ in true crown-ether fashion. *X*-Ray crystallography of a complex of *p-tert*-butylcalix[4]arene tetramethyl ether with Na$^+$ (*cf.* Figure 4.23) shows the Na$^+$ to be below the plane of the four oxygens, and this might also be true for the Li$^+$ calixarene monoanion. Possible structures for the tri- and tetra-anions are more difficult to suggest, and the apparent instability of the di-anion relative to its neighbors, the mono- and tri-anions, is a particularly puzzling phenomenon.

Extending these studies to two other Group-I cations, Gutsche and Nam treated a calix[4]arene with excess amounts of NaH and KH to generate the tetraanions in which Na^+ and K^+ are the counterions. The temperature-dependent ^1H NMR spectra of these solutions showed that the coalescence temperature is close to 160 °C for the Na^+ system and near room temperature for the K^+ system. Thus, the identity of the cation is seen to have a strong influence on the flexibility of the calix[4]arene tetraanions, the conformational stabilizing effect descending in the order $Li^+ > Na^+ > K^+$.

4.4.2 Calix[6]arene Oxyanions

Calix[6]arenes present a conformationally more convoluted system than the calix[4]arenes, and a study of their oxyanions has produced results from which the following conclusions have been drawn:[48] (a) The ability of a cation to reduce the conformational mobility of a calix[6]arene oxyanion increases with the size of the ion, *viz.* $Na^+ > K^+ > Rb^+ > Cs^+$, presumably because the larger cations better fill the 'lower rim' of the calixarene; (b) The inversion barrier of a calix[6]arene increases as the number of its anionic charges increases, *viz.* the T_c of the hexaanion is higher than that of the trianion, perhaps because of the forced approximation of negative charges that occurs during an inversion process proceeding *via* the 'OY through the annulus' pathway (see Figure 4.14); in the case of the oxyanions from calix[6]arene itself, however, the 'Ar through the annulus' pathway appears not to be precluded; (c) The pK_1, pK_2, and pK_3 values of calix[6]arenes are lower than the pK_a of acetic or carbonic acids, since treatment of DMSO solutions of calix[6]arenes with the acetates and carbonates of first- or second-group elements produces the trianions; (d) Calix[6]arene oxyanions are conformationally more stable than *p-tert*-butylcalix[6]arene oxyanions under almost all conditions; (e) Except in the case of Li^+, rapid exchange on the NMR timescale occurs between the 'bound' and 'unbound' forms of the metal ion in association with a calix[6]arene oxyanion in DMSO solution.

CHAPTER 5

Embroidering the Baskets: Reactions of Calixarenes and Introduction of Functional Groups

'Twirl follows twirl,
and every synthesis is the thesis of the next series'

Vladimir Nabokov, *Speak, Memory*, 1966

Much of the interest in the calixarenes derives from their promise as selective and useful complexation agents. This depends in part simply on the presence of cavities, but for the more sophisticated applications of complexation and catalysis it is also necessary that appropriate functional groups be present. This chapter discusses the various methods that have been devised for embroidering the calixarene baskets with functional groups.

5.1 Reactions at the 'Lower Rim' of the Calix

5.1.1 Ester Formation with Monofunctional Reagents

The 'lower rim' of phenol-derived calixarenes and the 'upper rim' of resorcinol-derived calixarenes are already functionalized with OH groups, and these provide excellent handles for affixing other moieties. The earliest reactions applied to the calixarenes, in fact, involved conversion of phenol-derived calixarenes into the acetates. The simple esters and ethers of the calixarenes are generally lower melting and more soluble than the parent compounds, and this can facilitate the measurements of properties such as cryoscopic molecular weight which was a datum of considerable importance in the early history of these compounds (see Chapter 1). Esterification can also be used effectively as a means for separating the components of a calixarene mixture. A good example is found in the work of Erdtman and Högberg who treated the crude reaction product obtained from *p*-bromobenzaldehyde and resorcinol with butyric anhydride and then separated the butyrates into a pair of pure isomers by fractional crystallization.[1,2]

---

[1] H. Erdtman and S. Högberg, *Tetrahedron Lett.*, **1968**, 1679.
[2] A. G. S. Högberg, *J. Am. Chem. Soc.*, **1980**, *102*, 6046.

When an excess of an acylating agent is used, generally all of the hydroxyl groups are converted into ester groups. With phenol-derived calix[4]arenes esterification freezes the compound into one or another of the conformers. Acetylation, for example, converts *p-tert*-butylcalix[4]arene mostly into the partial cone conformer, as discussed in Chapter 4. This is not necessarily true for all calixarenes and all acylating agents however, and a study has been carried out by Gutsche and coworkers in an attempt to discern the factors governing the conformational outcome of such reactions.[3] The systems chosen for investigation employ the $AlCl_3$-catalyzed and NaH-induced aroylation of calix[4]arene, *p-tert*-butylcalix[4]arene, and *p*-allylcalix[4]arene using benzoyl chlorides carrying various *para*-substituents. The ^{1}H NMR patterns arising from the methylene resonances can be used as a means for quite accurately determining the amount of cone, partial cone, and 1,3-alternate conformer (see Figure 5.1) that is present in the product (see Chapter 4).

In one set of experiments, using the $AlCl_3$ procedure, it was found that the product ratio of 1,3-alternate/partial cone conformer increases from 0.29 at 0 °C to 2.9 in refluxing chloroform. In another set of experiments, using the NaH procedure, the ratio of 1,3-alternate/cone conformer was shown to be strongly dependent on the *para*-substituent of the aroylating agent, as illustrated by the data in Table 5.1. With both *p-tert*-butylcalix[4]arene and *p*-allylcalix[4]arene, the ratio of cone/1,3-alternate conformer grows larger as the reactivity of the aroylating agent increases (as measured by its Hammett σ_{para} constant). The 'cross-over point' occurs much sooner in the sequence of aroylating agents with *p-tert*-butylcalix[4]arene than with *p*-allylcalix[4]arene however, and this correlates with the greater conformational mobility of the latter (see Table 4.1). These data have been interpreted in terms of a competition between conformational inversion and aroylation. Thus, lower temperatures (which decrease the rate of conformational interconversion) and more reactive aroylating agents (which compete more effectively with the conformational inversion) favor formation of the cone conformer. With a given aroylating agent the less conformationally mobile *p-tert*-butylcalix[4]arene is more likely to yield the cone conformer than the more conformationally mobile *p*-allylcalix[4]arene. The difference in the magnitude of the conformational inversion barriers is quite small, so the product composition must be the result of remarkably delicate balances between the rates of aroylation and conformational inversion.

Partial acetylation can occur under certain conditions. For example, a pair of acetates that are not conformational isomers can be isolated from the sulfuric-acid catalyzed reaction of acetic anhydride and *p-tert*-butylcalix[4]arene. One melts at 383—389 °C and is assigned the tetraacetate structure[4,5] **5** (R = *tert*-butyl, Y = $COCH_3$), and the other melts at 247—

---

[3] M. Iqbal, T. Mangiafico, and C. D. Gutsche, *Tetrahedron*, **1987**, *43*, 4917.

[4] V. Bocchi, D. Foina, A. Pochini, and R. Ungaro, *Tetrahedron*, **1982**, *38*, 373.

[5] C. D. Gutsche, B. Dhawan, J. A. Levine, K. H. No, and L. J. Bauer, *Tetrahedron*, **1983**, *39*, 409.

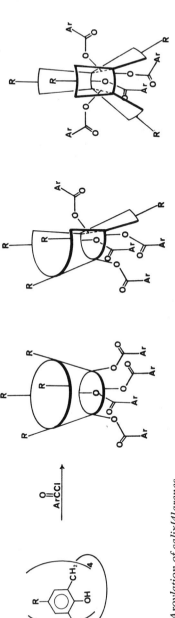

Figure 5.1 *Aroylation of calix[4]arenes*

Table 5.1 *Product composition in the aroylation of p-allylcalix[4]arene and p-tert-butylcalix[4]arene*[3]

Calixarene	R-C$_6$H$_4$COCl	σ_{para}	Product composition
p-Allyl	OCH$_3$	− 0.27	1,3-alternate
p-tert-Butyl			cone (90%), partial cone (5%), 1,3-alternate (5%)
p-Allyl	C(CH$_3$)$_3$	− 0.20	1,3-alternate, cone
p-tert-Butyl			cone (80%), 1,3-alternate (20%)
p-Allyl	CH$_3$	− 0.17	1,3-alternate
p-tert-Butyl			cone (90%), 1,3-alternate (5%), trisubstituted (5%)
p-Allyl	H	0.00	1,3-alternate
p-tert-Butyl			cone (90%), 1,3-alternate (trace)
p-Allyl	Br	0.23	1,3-alternate
p-tert-Butyl			cone
p-Allyl	CF$_3$	0.54	1,3-alternate (67%), cone (33%)
p-Allyl	CN	0.66	cone (major product), partial cone (minor product)
p-Allyl	NO$_2$	0.76	cone (95%), 1,3-alternate (5%)
p-tert-Butyl			cone

250 °C and is assigned the triacetate structure[5] **4** (R = *tert*-butyl, Y = COCH$_3$) (Figure 5.2). In cleaner fashion the AlCl$_3$-catalyzed benzoylation of calix[4]arene affords the tribenzoate **4** (R = H, Y = COC$_6$H$_5$) in excellent yield,[6] providing a material useful for 'upper rim' functionalization (see Figure 5.10). Also in surprisingly clean fashion, the AlCl$_3$-catalyzed reaction of calix[4]arenes with 3,5-dinitrobenzoyl chloride yields the 1,3-diester **3** (R = *tert*-butyl or H, Y = 3,5-dinitrobenzoyl) as the major product, a compound useful for introducing additional functional groups (see Figure 5.6).

The larger calixarenes generally behave like the calix[4]arenes. Thus, acid-catalyzed acetylations yield the hexaacetoxy and octaacetoxy compounds from calix[6]arenes and calix[8]arenes,[7] respectively. Benzoylation of calix[6]arene has been reported to yield the hexabenzoate[6] as a solid melting above 400 °C, but a reinvestigation of benzoyl chloride as well as a number of substituted benzoyl chlorides[8] shows that the product is mainly the 1,2,4,5-tetraester. With limiting amounts of benzoylating agent, two diesters of *p-tert*-butylcalix[6]arene have been isolated, one with the benzoyloxy groups in a 1,2 arrangement and the other in a 1,4 arrangement. The yields of these materials are quite low, however, limiting their synthetic utility.

5.1.2   Ether Formation with Monofunctional Reagents

Etherification, like esterification, can generally be effected on all of the OH groups of a calixarene regardless of ring size if sufficiently reactive reagents are used. A useful method for preparing the alkyl ethers involves treatment of the calixarene with an alkyl halide in THF–DMF solution in the presence of sodium hydride. Methyl, ethyl, allyl, and benzyl ethers have all been prepared in high yields by this method.[5] A series of polyalkyloxy ethers were synthesized in the early work on calixarenes by Cornforth and coworkers[9] who used the tosylate of the alkylating agent in the presence of potassium *tert*-butoxide. To insure that all of the hydroxyl groups of the calixarene had reacted they followed the reaction by means of UV spectroscopy, noting the disappearance of the absorption at *ca.* 300 nm and its replacement by a pair of peaks in the 270—280 nm region. The use of tosylates in other instances, however, has led to partial reaction. Thus, treatment of *p*-allylcalixarene with benzyl tosylate yields the dibenzyl ether[5] **3** (R = allyl, Y = benzyl). The action of diazomethane on *p-tert*-butylcalixarene also yields a diether **3** (R = *tert*-butyl, Y = Me), although a crystal obtained from the product was found by *X*-ray determination to be the monomethyl ether **2** (R = *tert*-butyl, Y = Me).[10] *p-tert*-Butylcalix[4]arene treated with dimethyl sulfate in the presence of BaO–Ba(OH)$_2$ in DMF yields the trimethyl ether[5] **4** (R = *tert*-butyl, Y = Me).

[6] C. D. Gutsche and L.-g. Lin, *Tetrahedron*, **1986**, *42*, 1633.
[7] C. D. Gutsche, B. Dhawan, K. H. No, and R. Muthukrishnan, *J. Am. Chem. Soc.*, **1981**, *103*, 3782.
[8] C. D. Gutsche, J. S. Rogers, and C. C. Johnson, to be published.
[9] J. W. Cornforth, E. D. Morgan, K. T. Potts, and R. J. W. Rees, *Tetrahedron*, **1973**, *29*, 1659.
[10] C. D. Gutsche and G. F. Stanley, unpublished work.

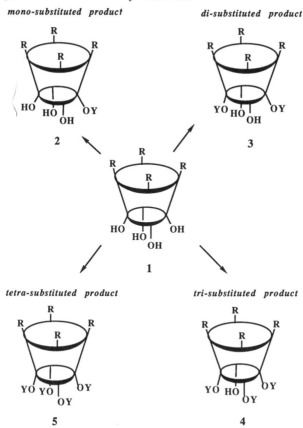

Figure 5.2 *Esterification and etherification pathways for calix[4]arenes*

The trimethylsilyl ethers of the calixarenes provide useful derivatives. The calix[6]arenes and calix[8]arenes react with standard reagents such as hexamethyldisilazene and chlorotrimethylsilane to form the corresponding hexa- and octa-trimethylsilyl ethers. Those from the calix[4]arenes, however, are more difficult to prepare and generally require especially reactive reagents such as N,O-bis(trimethylsilyl)acetamide.

5.1.3 Esterification and Etherification with Polyfunctional Reagents

Through the use of appropriately functionalized esterification and etherification reagents additional and/or new functional groups can be introduced into the calixarenes. Among the first of the examples in this category was the preparation of the 2,4-dinitrophenyl ethers of *p-tert*-butylcalix[8]arene[11] (**8**)

[11] R. Muthukrishnan and C. D. Gutsche, *J. Org. Chem.*, **1979**, *44*, 3962.

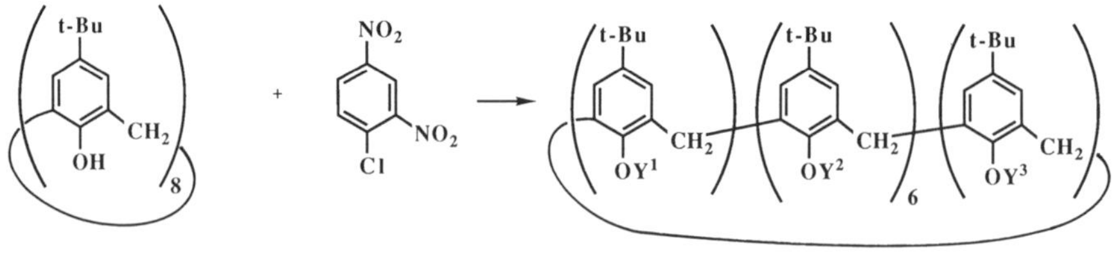

Figure 5.3   *2,4-Dinitrophenyl ethers of* p-tert-*butylcalix[8]arene*[11]

obtained by reaction of **6** with 2,4-dinitrochlorobenzene (**7**) in pyridine solution, as illustrated in Figure 5.3. Depending on the ratio of calixarene to arylating agent, the product is **8** (Y^1 = 2,4-dinitrophenyl, Y^2 = Y^3 = H), or **8** (Y^1 = Y^3 = 2,4-dinitrophenyl, Y^2 = H), or **8** (Y^1 = Y^3 = H, Y^2 = 2,4-dinitrophenyl). Even with an excess of arylating agent two hydroxyl groups are not arylated, although they can be acetylated by treatment of the hexaaryl compound with acetyl chloride to produce **8** (Y^1 = Y^3 = $COCH_3$, Y^2 = 2,4-dinitrophenyl). As mentioned in Chapter 3 it was the seemingly inexplicable behavior of 2,4-dinitrochlorobenzene with what was thought at the time to be the cyclic tetramer that provided the earliest clue that the material was actually the cyclic octamer.

The first water-soluble calixarene was prepared by Ungaro and colleagues[12] by treating *p-tert*-butylcalix[4]arene with *tert*-butyl α-bromoacetate and NaH in THF solution to give a 70% yield of the tetra-ester **9** (R = *tert*-butyl) (see Figure 5.4). Base-induced hydrolysis produces the corresponding tetraacid **9** (R = H), soluble in water to the extent of *ca.* 10^{-4} M. Using potassium *tert*-butoxide rather than NaH as the base, the diester **10** (R = *tert*-butyl) is produced in 40% yield, and from it the corresponding diacid **10** (R = H) can be obtained by hydrolysis.[13] In a further elaboration of this chemistry, the Parma group carried out this same procedure on a 'lower rim' bridged calixarene to produce the crown-5-diacetic acid compound **11**[13] shown in Figure 5.4.

Following the precedent set by the Parma laboratories, the theme of using ether formation to introduce functionality has been exploited by several other workers. Anthony McKervey and his research group at the University of Cork[14] have prepared an entire series of compounds containing the various structural combinations **12—14** (**c—f**) shown in Figure 5.5, starting with the *p-tert*-butyl calixarenes **12a**, **13a**, and **14a** as well as the *p*-H calixarenes **12b**, **13b**, and **14b**. This same research group provided additional examples from the cyclic tetramers, *viz.* **12** (R^1 = CO_2Et, R^2 = *tert*-butyl),[15] **12**

[12] A. Arduini, A. Pochini, S. Reverberi, and R. Ungaro, *J. Chem. Soc., Chem. Commun.*, **1984**, 981.
[13] R. Ungaro, A. Pochini, and G. D. Andreetti, *J. Inclusion Phenom.*, **1984**, *2*, 199.
[14] M. A. McKervey, E. M. Seward, G. Ferguson, B. Ruhl, and S. J. Harris, *J. Chem. Soc., Chem. Commun.*, **1985**, 388.
[15] M. A. McKervey and E. M. Seward, *J. Org. Chem.*, **1986**, *51*, 3581.

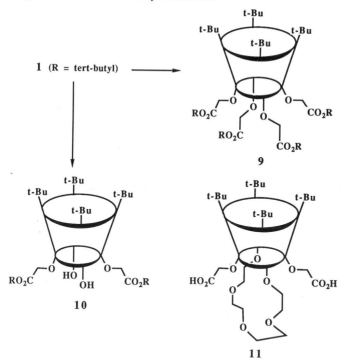

Figure 5.4   *Water-soluble 'lower rim' functionalized calixarenes*

($R^1 = CH_2COCH_3$, $R^2 = tert$-butyl), and **12** ($R^1 = CH_2COCH_3$, $R^2 = H$),[16] and Chang and coworkers[17] have made the amides **12g**, **13g**, and **14g**. Commercial interest in these compounds is indicated by several patents, as discussed in Chapter 7.

The reaction of calix[4]arenes with 3,5-dinitrobenzoyl chloride in the presence of $AlCl_3$ to form the diesters has been used by Gutsche and coworkers[18] to construct 'lower rim' functionalized calixarenes in the manner illustrated in Figure 5.6. Reduction of the four nitro-groups of **15** yields the tetraamino compound **16** which is treated with difunctional spanning reagents to afford compounds of structure **17** that have been named 'double-cavity calixarenes'. The spanner that works best is adipoyl chloride, which produces **17** ($R = tert$-butyl, $X = H$, $n = 4$) in *ca.* 40% yield. In addition to the bis(3,5-dinitrobenzoyl) esters, the bis(3,5-dinitro-4-methylbenzoyl) ester (**15**, $X = Me$), the bis(3,5-dinitro-4-dimethylaminomethyl) ester (**15**, $X = Me_2NCH_2$), and the bis(3,5-dinitro-4-vinyl) esters (**15**, $X = CH=CH_2$) have been prepared.[19]

[16] G. Ferguson, B. Kaitner, M. A. McKervey, and E. M. Seward, *J. Chem. Soc., Chem. Commun.*, **1987**, 584.

[17] S.-K. Chang, S.-K. Kwon, and I. Cho, *Chem. Lett.*, **1987**, 947.

[18] C. D. Gutsche, M. Iqbal, K. C. Nam, K. A. See, and I. Alam, *Pure Appl. Chem.*, **1988**, *60*, 483.

[19] C. D. Gutsche and K. A. See, to be published.

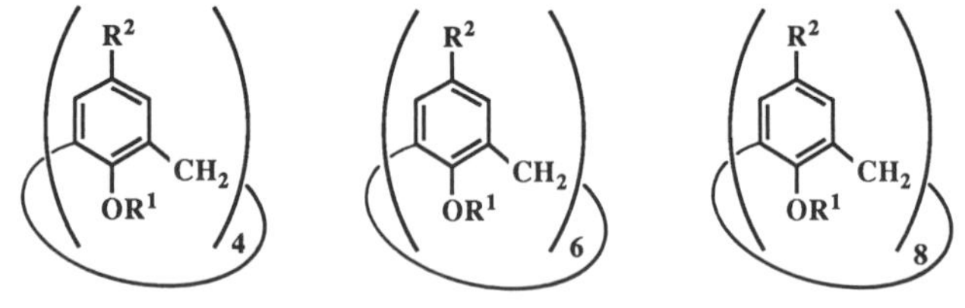

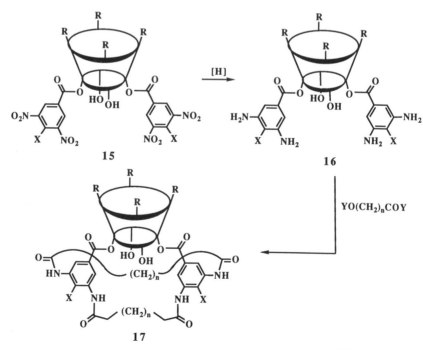

a	R^1 = H, R^2 = *tert*-butyl
b	R^1 = H, R^2 = H
c	R^1 = CH_2CO_2Et, R^2 = *tert*-butyl
d	R^1 = CH_2CO_2Me, R^2 = *tert*-butyl
e	R^1 = CH_2CO_2Et, R^2 = H
f	R^1 = CH_2CO_2Me, R^2 = H
g	R^1= $CH_2CONHBu$, R^2 =*tert*-butyl

Figure 5.5   *'Lower   rim'   functionalized   calix[4]arenes,   calix[6]arenes,   and calix[8]arenes*[14-17]

Figure 5.6   *'Lower rim' functionalized 'double cavity' calixarenes*[18,19]

5.2   Reactions at the 'Upper Rim' of the Calix

5.2.1   Phenol-derived Calixarenes

5.2.1.1   Dealkylation of p-*Alkylcalixarenes.* It is a fortunate circumstance that the few phenols that provide calixarenes in good yields in the one-step procedures, *viz.* p-*tert*-butylphenol, *p-tert*-pentylphenol, and *p*-(1,1,3,3-tetramethylbutyl)phenol, are those from which the *para*-substituent can be easily removed by a reverse Friedel–Crafts reaction. The Lewis acid-catalyzed transalkylation, known for many years, was applied in 1975 to phenols by Tashiro and coworkers.[20] The Mainz group,[21] using a similar procedure, noted that, while it failed to work very well in removing *tert*-butyl groups from linear tetranuclear phenol–formaldehyde oligomers, it gave reasonable yields with the cyclic analogs. The experimental details were given in a subsequent paper[22] which described the selective removal of the *tert*-butyl groups from dimethyl-di-*tert*-butylcalix[4]arene to produce dimethyl-calix[4]arene in 45% yield. The procedure is now extensively used as a means for obtaining calixarenes containing unsubstituted *para*-positions, thus making them amenable to functionalization. As illustrated in Figure 5.7 by the conversion of **18** into **19**, calix[4]arene, for example, is prepared in 65–70% yield,[23,24] calix[6]arene in 89% yield,[6] and calix[8]arene in 93% yield,[6] the latter two reactions being conducted in toluene containing some phenol as a better acceptor molecule for the *tert*-butyl group. Selective removal of *tert*-butyl groups can sometimes be effected, those *para* to a free OH group being released more readily than those *para* to an esterified OH group. An example is the transformation of compound **20** to **21** in 80–90% yield by treatment with a ten-fold excess of AlCl$_3$ in toluene to give a compound useful for selective introduction of functional groups.

5.2.1.2   Electrophilic Substitution. Although the most obvious method for functionalizing a *p*-H calixarene uses aromatic electrophilic substitution reactions, this procedure has enjoyed success only recently. Early attempts by the St Louis group to effect nitration of calix[4]arene gave a product that was difficult to purify, and attention was turned to the other methods of functionalization that are described in the next sections of this chapter. In the hands of Seiji Shinkai of Kyushu University, however, electrophilic sustitution has been shown to work very well. Shortly after the announcement by the Parma group of a water-soluble calixarene (see p. 76–7, 132) a short communication by Shinkai and coworkers,[25] followed two years later by a longer paper,[26]

[20] M. Tashiro, G. Fukata, S. Mataka, and K. Oe, *Org. Prep. Proced. Int.*, **1975**, *7*, 231. See M. Tashiro, *Synthesis*, **1979**, 921 for general references.

[21] V. Böhmer, D. Rathay, and H. Kämmerer, *Org. Prep. Proced. Int.*, **1978**, *10*, 113.

[22] H. Kämmerer, G. Happel, V. Böhmer, and D. Rathay, *Monatsh.*, **1978**, *109*, 767.

[23] C. D. Gutsche and J. A. Levine, *J. Am. Chem. Soc.*, **1982**, *104*, 2652.

[24] C. D. Gutsche, J. A. Levine, and P. K. Sujeeth, *J. Org. Chem.*, **1985**, *50*, 5802.

[25] S. Shinkai, S. Mori, T. Tsubaki, T. Sone, and O. Manabe, *Tetrahedron Lett.*, **1984**, *25*, 5315.

[26] S. Shinkai, K. Araki, T. Tsubaki, T. Arimura, and O. Manabe, *J. Chem. Soc., Perkin Trans. 1*, **1987**, 2297.

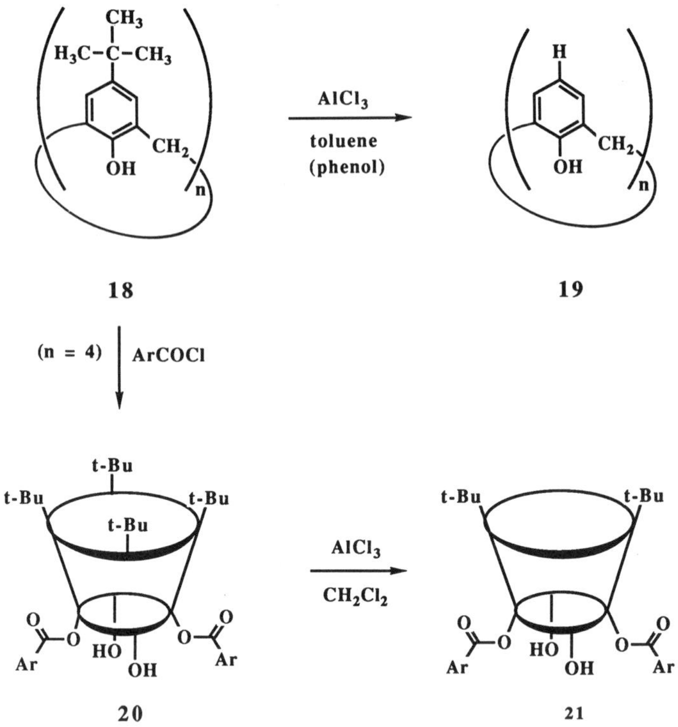

Figure 5.7   *De-*tert*-butylation of calixarenes*

described the conversion of calix[6]arene (**19**, $n = 6$) to the quite water-soluble *p*-sulfonatocalix[6]arene (**22**, $n = 6$) in 75% yield by treatment with sulfuric acid at 100 °C. Subsequently, they improved the synthesis by eliding the removal of the *tert*-butyl groups and the addition of the sulfonate groups into a single step, converting *p-tert*-butylcalix[6]arene directly into *p*-sulfonatocalix[6]arene in 50% yield by careful control of the time and temperature.[27] It is probable that a similar de-*tert*-butylation and electrophilic substitution was unwittingly achieved by Zinke who reported the obtention of an explosive compound by treatment of a *p-tert*-butylcalixarene (ring size uncertain) with nitric acid.[28]

Using *p*-sulfonatocalix[6]arene as the starting material, Shinkai *et al.* made the corresponding *p*-nitro compound (**23**, $n = 6$) by treatment with nitric acid for 10 h at 0—5 °C which afforded a 15% yield of *p*-nitrocalix[6]arene[29] (see Figure 5.8). The general utility of these methods was demonstrated by extending them to the calix[4]arene and calix[8]arene, the sulfonato derivatives being obtained in yields of 78—88% and the nitro derivatives in yields of *ca.* 20%. Then, by using the chemistry described in the previous section, a

[27] S. Shinkai, H. Kawaguchi, and O. Manabe, *J. Polymer Sci., Polymer Lett.*, **1988**, *26*, 391.
[28] A. Zinke, R. Ott, and H. Garrana, *Monatsh.*, **1958**, *89*, 135.
[29] S. Shinkai, T. Tsubaki, T. Sone, and O. Manabe, *Tetrahedron Lett.*, **1985**, *26*, 3343.

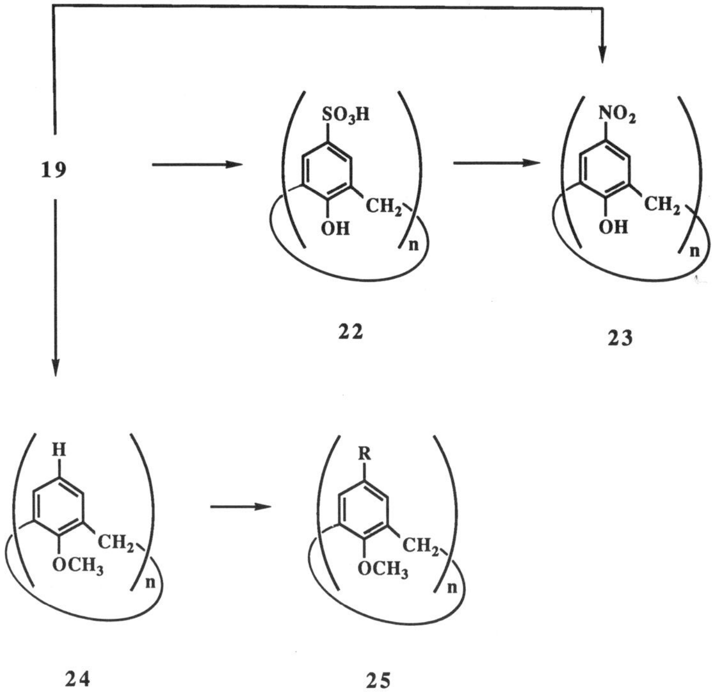

Figure 5.8 *Electrophilic substitution reactions of calixarenes*

calixarene functionalized at both the 'upper' and 'lower' rims (**26**) was prepared.[30] An attempt by Shinkai to achieve direct nitration of calix[6]arene[29] resulted in oxidation of the starting material as evidenced by the appearance of carbonyl stretching absorptions in the infrared spectrum of the crude product. Although it is reported in a later paper[26] that a 27% yield of *p*-nitrocalix[8]arene can be isolated by direct nitration, the preference of these authors remains with the indirect route *via* sulfonation. Subsequent studies in the St Louis laboratories[31] have confirmed that it is, in fact, pos-

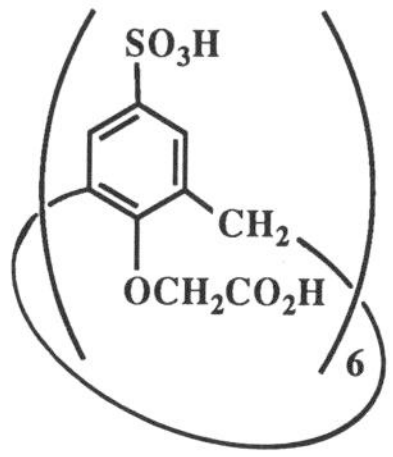

26

[30] S. Shinkai, H. Koreishi, K. Ueda, and O. Manabe, *J. Chem. Soc., Chem. Commun.*, **1986**, 233.
[31] K. C. Nam, Ph. D. Thesis, Washington University, St Louis, **1987**.

sible to obtain *p*-nitrocalix[4]arene by direct nitration but that purification severely reduces the yield. These difficulties appear to have been overcome in the Korean laboratory of Kwanghyun No which reports[32] that *p*-nitrocalix[4]arene can be obtained (a) in 87% yield by treatment of calix[4]arene with nitric acid in acetic acid–benzene solution and (b) in 53% yield by treatment with NaOH and NaNO$_3$ in water followed by oxidation with dilute nitric acid.

Acetylation and benzoylation of calix[4]arene under Friedel–Crafts conditions, using acetyl chloride or benzoyl chloride and AlCl$_3$, yields only O-substituted products,[6] and the esters that are formed fail to undergo further reaction at the *p*-positions. On the basis of the Hammett constants for the OCOCH$_3$ group (σ_{para} = + 0.31) and the OCH$_3$ group (σ_{para} = − 0.27) it was postulated that the calixarene ethers might be more likely to undergo *p*-acylation and aroylation, and this proved to be true.[6] Unfortunately, the process is complicated in the case of the calix[4]arene methyl ether by concomitant demethylation that produces a mixture requiring separation by column chromatography, a procedure generally inimical to large-scale synthetic operation. In contrast, the methyl ethers of calix[6]arene and calix[8]arene afford 60—65% yields of easily purified *p*-acetylcalix[6]arene (**25**, *n* = 6, R = COCH$_3$) and *p*-acetylcalix[8]arene (**25**, *n* = 8, R = COCH$_3$) (Figure 5.8), respectively.[6] Haloform reactions on these acetyl compounds produce the corresponding *p*-carboxycalix[6]arene (**25**, *n* = 6, R = CO$_2$H) and *p*-carboxycalix[8]arene (**25**, *n* = 8, R = CO$_2$H), providing further examples of calixarenes carrying water solubilizing groups (see Figure 5.8). Although early attempts to effect a Fries rearrangement of the calixarene acetates failed, No and coworkers[33] have succeeded in finding appropriate conditions and report that a 64% yield of *p*-acetylcalix[4]arene can be achieved by heating the tetraacetate with AlCl$_3$ for 18 h at 150 °C.

With the realization that the calixarene ethers are better behaved in electrophilic substitution reactions than are the parent calixarenes, other such reactions were explored.[34] Among these was bromination which, with N-bromosuccinimide, proceeds with calix[4]arene methyl ether in 90% yield to give **25** (*n* = 4, R = Br). This, in turn, led to *p*-cyanocalix[4]arene methyl ether (**25**, *n* = 4, R = CN) by treatment with CuCN, *p*-lithiocalix[4]arene methyl ether (**25**, *n* = 4, R = Li) by treatment with *n*-butyllithium, and *p*-carboxycalix[4]arene methyl ether (**25**, *n* = 4, R = CO$_2$H) by carbonation of the lithio compound.[34]

Special attention has been given to the introduction of aryl moieties into the *p*-positions of a calix[4]arene because of the deep cavity that would be created. Since arylation of *p*-lithiocalix[4]arene methyl ether failed,[34] benzoylation was resorted to as a means for achieving this end while recognizing that the shapes of *p*-benzoyl and *p*-phenylcalix[4]arenes are quite different (see Figure 4.4). Treatment of calix[4]arene methyl ether with benzoyl chloride or

[32] K. No and Y. Noh, *Bull. Korean Chem. Soc.*, **1986**, *7*, 314.
[33] K. No, Y. Noh, and Y. Kim, *Bull. Korean Chem. Soc.*, **1986**, *7*, 442.
[34] C. D. Gutsche and P. F. Pagoria, *J. Org. Chem.*, **1985**, *50*, 5795.

27

28

p-methoxybenzoyl chloride (anisoyl chloride) leads to 27 (R = Me, Y = H) and 27 (R = Me, Y = OMe) in 32 and 58% yield from which the ether groups can be removed, though with difficulty, by refluxing with EtSNa in DMF.[34]

An interesting facet of the acylation reaction is revealed by experiments with the 1,3-bis(3,5-dinitrobenzoyl)calix[4]arene 21 (Figure 5.7) which undergoes acylation at the two available *para*-positions to yield 28 rather than forming a tetraester as is observed with calix[4]arene (19, *n* = 4) (Figure 5.7). This is probably the result of conformational differences between the molecules, 18 preferring the cone form while 21 has been shown to exist in the solid state in the 1,3-alternate form[35] in which the free OH groups at C-25 and C-27 are shielded by the *tert*-butyl groups at C-5 and C-17. Among the compounds prepared in this fashion is one carrying acryloyl moieties (28, R = HC=CH₂), an interesting candidate for further functionalization by alterations in the double bond of the R groups.

5.2.1.3 Functionalization via para-*Claisen Rearrangement.* The initial difficulties encountered with the electrophilic substitution reactions of calixarenes, particularly in the calix[4]arene series, prompted Gutsche and coworkers to explore alternative routes. One of the most productive of these employs the *para*-Claisen rearrangement, as illustrated in Figure 5.9.[23,24] Calix[4]arene (19, *n* = 4) is readily converted into the tetraallyl ether 29 (*n* = 4), and when this material is heated in refluxing N,N-diethylaniline it undergoes a *para*-Claisen rearrangement to produce *p*-allylcalix[4]arene (30, *n* = 4) in *ca.* 75% yield. The *p*-toluenesulfonate of this product (31, R = allyl) has been converted into a variety of functionalized calix[4]arenes, including the *p*-(2-oxoethyl) compound 31a, the *p*-(2-hydroxyethyl) compound 31b, the *p*-(2-bromoethyl) compound 31c, the *p*-(2-azidoethyl) compound 31d, the *p*-(2-aminoethyl) compound 31e, the *p*-(2-cyanoethyl) compound 31f, the *p*-formyl compound 31g, and the oxime 31h. Removal of the tosyl group can be effected with refluxing aqueous alcoholic base as illustrated, for example, by the conversion of 31e into 32e — a compound which, along with 31e, is discussed in Chapter 6 with respect to its interaction with metal ions.

[35] C. D. Gutsche, K. A. See, and F. R. Fronczek, unpublished work.

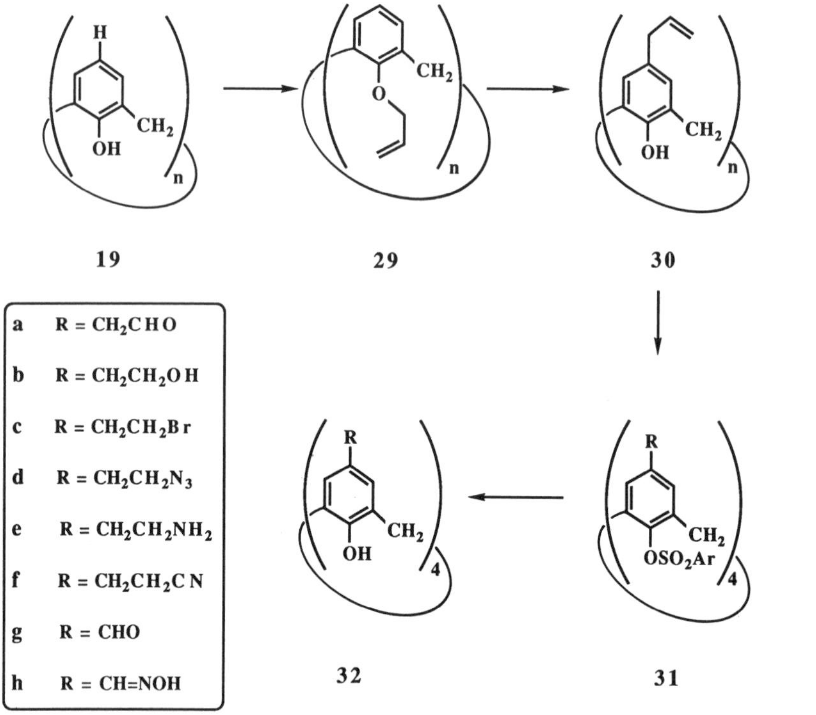
H
CH2
OH
n
19
O
CH2
n
29
CH2
OH
n
30
R
CH2
OSO2Ar
4
31
R
CH2
OH
4
32
a R = CH2CHO
b R = CH2CH2OH
c R = CH2CH2Br
d R = CH2CH2N3
e R = CH2CH2NH2
f R = CH2CH2CN
g R = CHO
h R = CH=NOH

Figure 5.9 *Functionalization of calixarenes* via *the* para-*Claisen rearrangement route*[23,24]

The *para*-Claisen rearrangement route should also be applicable to the larger calixarenes, but the results are somewhat disappointing. The hexaallyl ether of calix[6]arene (**29**, $n=6$) is readily prepared, and it undergoes a six-fold *para*-Claisen rearrangement to afford *p*-allylcalix[6]arene (**30**, $n=6$). Although the crude product shows spectral evidence for large amounts of the desired product, only 21% of pure material has been obtained from the reaction mixture.[6] The octaallyl ether of calix[8]arene (**29**, $n=8$) is also prepared in good yield,[6] but attempts to isolate pure *p*-allylcalix[8]arene from the *para*-Claisen rearrangement have been unsuccessful.

The facile synthesis of the tribenzoate of calix[4]arene (**4**, R=H, Y=COC_6H_5) as described in Section 5.1.1 (see Figure 5.2) provides a method[6] for preparing monoallylcalix[4]arene *via* the *para*-Claisen rearrangement route. Conversion of the tribenzoate into 25-allyloxy-26,27,28-tribenzoyloxy[4]arene (**33**) in 58% yield followed by refluxing in N,N-diethylaniline produces 5-allyl-25-hydroxy-26,27,28-tribenzoyloxy-calix[4]arene (**34**) in 62% yield, as shown in Figure 5.10. Hydrolysis in aqueous alcoholic NaOH affords the monoallylcalix[4]arene (**35**) in 72.5% yield. Alternatively, the benzoyl groups can be removed prior to the Claisen rearrangement, producing 25-allyloxy-26,27,28-trihydroxycalix[4]arene (**36**) which rearranges to yield **35**. The former route is preferred.

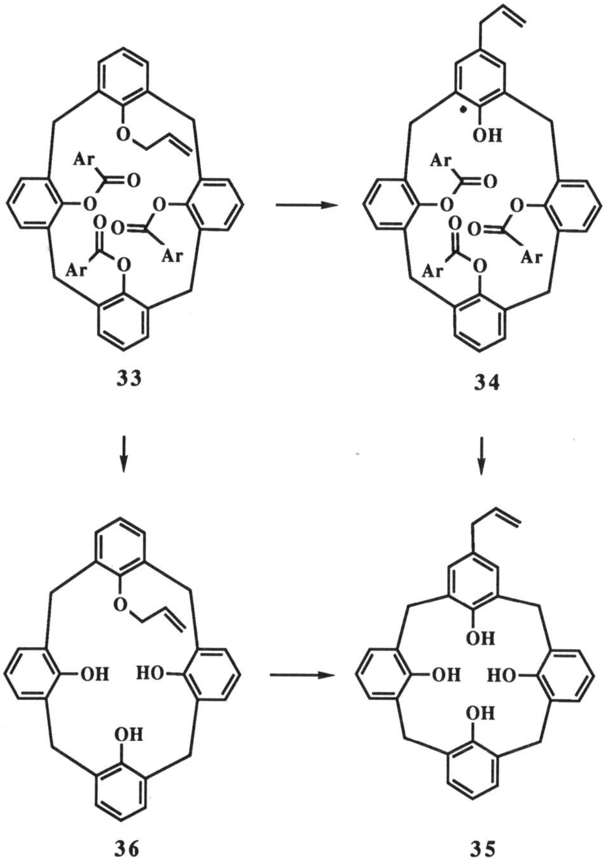

Figure 5.10 *Synthesis of the monoallylcalix[4]arene*[6]

5.2.1.4 Functionalization via *the* p-*Quinonemethide Route.* Another altern-ative to the direct electrophilic substitution route takes advantage of the nucleophilic character of the *para*-position of phenolates.[36] When calix[4]arene is mixed with formaldehyde along with any of a number of secondary amines a Mannich-type reaction occurs to produce *p*-alkyl-aminomethylcalix[4]arenes (**37a–h**), as illustrated in Figure 5.11. These compounds are functionalized calixarenes in their own right and, as will be discussed in the next chapter, are sufficiently soluble in aqueous acid to be useful in complexation studies. In addition, they also serve as intermediates for a functionalization pathway in which the aminomethylcalix[4]arenes (**37**) are converted into the corresponding quaternary ammonium compounds (**38**) and these, in turn, are treated with two equivalents of a nucleophile to afford calix[4]arenes of the general structure **40**. Although *p*-quinone-methides (**39**) have not been isolated or directly detected as intermediates,

[36] C. D. Gutsche and K. C. Nam, *J. Am. Chem. Soc.*, **1988**, *110*, 6153.

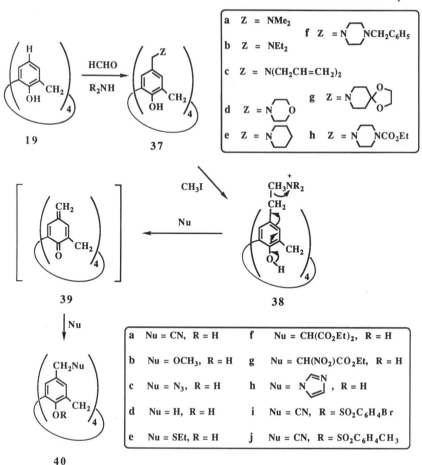

Figure 5.11 *Functionalization of calix[4]arenes* via *the* p-*quinonemethide route*[36]

there is strong presumptive evidence to view them as the probable precursors to the products that are isolated. The *p*-quinonemethide route provides a particularly short pathway to a variety of functionalized calixarenes. For example, the conversion of *p-tert*-butylcalix[4]arene into *p*-(2-amino-ethyl)calix[4]arene (**32e**) *via* the *para*-Claisen rearrangement route is an 8-step sequence, as pictured in Figure 5.9, whereas only five steps are required in the *p*-quinonemethide sequence which involves debutylation, preparation of the Mannich base, quaternization, treatment with NaCN to yield the nitrile **40a**, and reduction of the nitrile to the amine **32e**. The *p*-quinonemethide route also provides an easy method for introducing carboxyl groups onto the calixarene framework, enolates of diethyl malonate and ethyl α-nitroacetate reacting smoothly with the quaternary compounds to yield **40f** and **40g**, respectively. Hydrolysis and decarboxylation of **40f** produces *p*-(carboxyethyl)calix[4]arene (**42**, *n* = 4), the first member of a series of calixarenes that includes *p*-carboxyethylcalix[5]arene (**42**, *n* = 5), *p*-carboxyethylcalix[6]arene (**42**, *n* = 6), *p*-carboxyethylcalix[7]arene (**42**, *n* = 7), and *p*-carboxyethylcalix[8]arene (**42**, *n* = 8) (see Figure 5.12). All of these compounds are available in good yield by the *p*-quinonemethide procedure using the Mannich bases **41** (*n* = 4—8) as intermediates.[37]

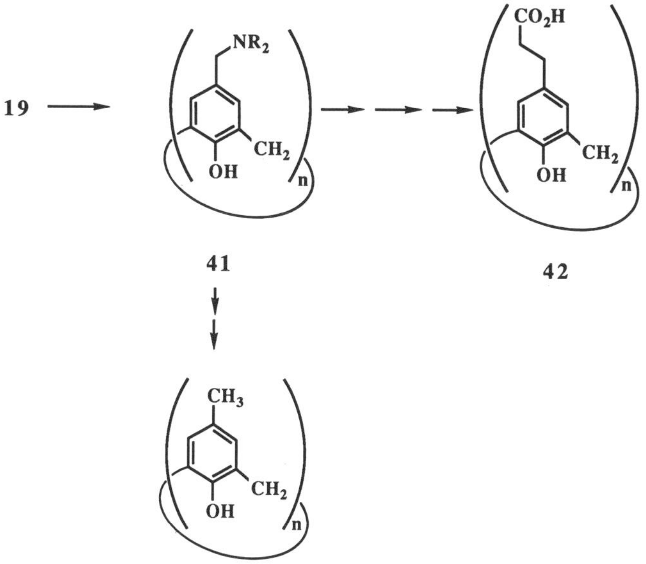

Figure 5.12 *Synthesis of aminocalixarenes and carboxycalixarenes by the* p-*quinonemethide route*[36]

The *p*-quinonemethide route, though remarkably effective in providing a way to introduce certain functional groups onto the calixarenes, is not without shortcomings. Attempts to effect nucleophilic substitution of the quaternized Mannich bases with acetylides, for example, have failed, probably because of competition from the anion of the solvent, DMSO. Nucleophiles that are too weakly basic also fail to effect a reaction, the limit appearing to be in the vicinity of imidazole, which reacts slowly to give the interesting tetraimidazolylmethyl compound **40h**.

The reaction of **38** with sodium borohydride to yield *p*-methyl-calix[4]arene (**43**, *n* = 4), is only modestly successful, the reaction being complicated by hydride displacement on the methyl groups of the quaternary salt to yield the parent amine **37a**. Nevertheless, sufficient quantities of **43** were isolated to allow its purification and characterization. This is the compound that was reported thirty years earlier by Hayes and Hunter as the product of

[37] C. D. Gutsche and I. Alam, *Tetrahedron*, **1988**, *44*, 4689.

their 'rational synthesis' (see Figure 1.1) — the compound that was tacitly assumed to provide a proof of structure for the Zinke products. The conversion of **19** into **43** provides the connection between the one-step and multistep syntheses, as discussed in Chapter 3 (see Figure 3.4).

5.2.1.5   Functionalization via *the* p-*Chloromethylation Route.* The embroidering of calixarenes *via* functionalization has been further expanded by a *p*-chloromethylation route introduced by the Parma group,[38] illustrated in Figure 5.13. Treatment of **19** with octyl chloromethyl ether and $SnCl_4$ affords an 80% yield of *p*-chloromethylcalix[4]arene (**44**, Y=H, *n*=4). From **44** a variety of compounds have been obtained, including *p*-methylcalix[4]arene (**45a**) *via* $LiAlH_4$ (83%), *p*-ethylcalix[4]arene (**45b**) *via* MeLi (35%), *p*-benzylcalix[4]arene (**45c**) *via* benzene in the presence of BF_3 (40%), *p*-mesitylmethylcalix[4]arene (**45d**) *via* mesitylene in the presence of BF_3 (60%), *p*-(3,5-dimethyl-4-hydroxybenzyl)calix[4]arene (**45e**) *via* 2,6-dimethylphenol

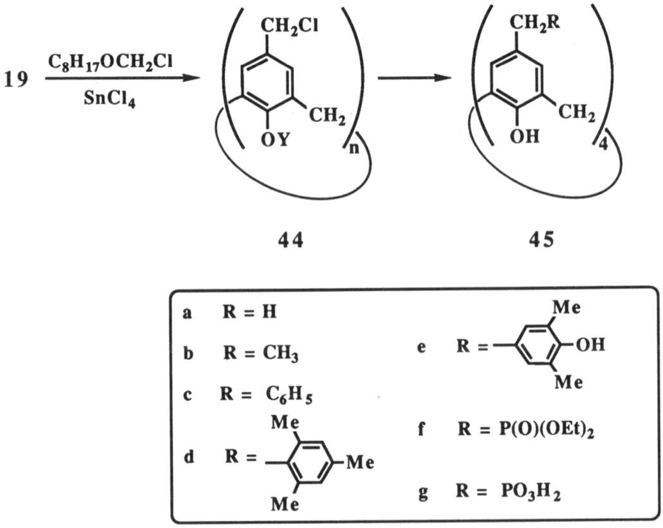

Figure 5.13   *Functionalization of calix[4]arenes* via *the* p-*chloromethylation route*[38]

and BF_3, and *p*-diethylphosphorylmethylcalix[4]arene (**45f**) *via* triethylphosphite (>90%). Hydrolysis of **45f** yields the corresponding phosphate **45g**. In a comparable fashion the methyl ethers of *p-tert*-butylcalix[6]arene and *p-tert*-butylcalix[8]arene have been treated with dimethyoxymethane and $ZnCl_2$ to produce the corresponding *p*-chloromethyl compounds **44** (Y=Me, *n*=6 and 8) in 85% yield.

[38] A. Arduini, A. Casnati, A. Pochini, and R. Ungaro, Fifth International Symposium on Inclusion Phenomena and Molecular Recognition, Orange Beach, Ala, **1988**, p. H13.

5.2.2 Resorcinol-derived Calixarenes

The resorcinol-derived calixarenes carry eight hydroxyl groups just below the 'upper rim' of the calix. The manner in which bridges have been built between these groups is discussed in Chapter 4, and attention is now turned to the ways that these bridged calix[4]resorcinarenes have been functionalized and transformed into closed baskets by the elegant chemistry of Cram and his coworkers.[39-42] Starting with the compound **46**, obtained by acid-catalyzed

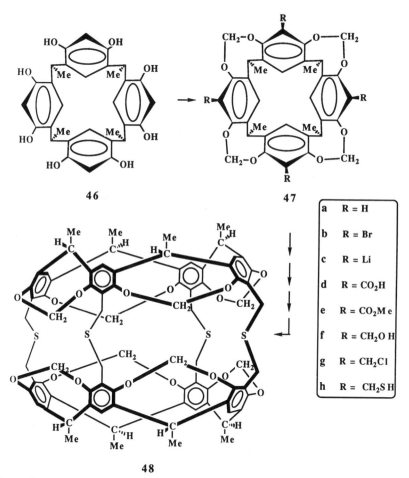

a	R = H
b	R = Br
c	R = Li
d	R = CO₂H
e	R = CO₂Me
f	R = CH₂OH
g	R = CH₂Cl
h	R = CH₂SH

Figure 5.14 *Synthesis of a carcerand from a resorcinol-derived calixarene*[39-42]

[39] R. Moran, S. Karbach, and D. J. Cram, *J. Am. Chem. Soc.*, **1982**, *104*, 5826.

[40] D. J. Cram, S. Karbach, Y. H. Kim, L. Baczynskyj, and G. W. Kalleymeyn, *J. Am. Chem. Soc.*, **1985**, *107*, 2575.

[41] D. J. Cram, S. Karbach, H. E. Kim, C. B. Knobbler, E. F. Maverick, J. L. Ericson, and R. C. Helgeson, *J. Am. Chem. Soc.*, **1988**, *110*, 2229.

[42] D. J. Cram, S. Karbach, Y. H. Kim, L. Baczynskyj, K. Marti, R. M. Sampson, and G. W. Kalleymeyn, *J. Am. Chem. Soc.*, **1988**, *110*, 2554.

condensation of resorcinol and acetaldehyde (see Section 2.2.2), methylene bridges are introduced to yield **47a**. Treatment with N-bromosuccinimide introduces bromines in the positions between the oxygen functions of the aromatic rings to give **47b** which is lithiated to give **47c**, carbonated with CO_2 to give **45d**, and esterified with diazomethane to give **47e**. Lithium aluminum hydride reduction of **47d** leads to the alcohol **47f** from which the two units necessary for the final construction are produced, *viz.* the tetrachloro compound **47g** [by treatment of **47f** with $(C_6H_5)_3P$ and N-chlorosuccinimide] and the tetrathiol compound **47h** (by treatment of **47g** with thiourea followed by hydrolysis). The critical step of joining **47g** and **47h** is carried out under high dilution conditions in a DMF–dioxane solution using $CsCO_3$ as the base. The product, obtained in *ca.* 29% yield, is a gray-white powder whose exotic complexation capacities are commensurate with its designation as a 'carcerand', as discussed in the next chapter.

5.3   Oxidation Reactions of Calixarenes

5.3.1   Oxidation of the Methylene Bridges

Moshfegh and coworkers[43] have reported that the action of chromic acid on calixarene esters **49**($n = 4$) converts the methylene groups into keto functions (**50**) (Figure 5.15), although the products seem not to be very well characterized. Ninagawa and coworkers[44] have made similar observations with the cyclic tetramers as well as with the cyclic hexamers and cyclic octamers. The reactions of these compounds, carried out with chromium trioxide in a mixture of acetic anhydride and acetic acid, yielded products (**51**) in which a single methylene group has been oxidized in the case of *p-tert*-butylcalix[4]arene (**51a**; $x = 3$, $y = 1$) and *p-tert*-butylcalix[6]arene (**51a**; $x = 5$, $y = 1$) and three methylene groups have been oxidized in the case of *p-tert*-butylcalix[8]arene (**51a**; $x = 5$, $y = 3$). Hydrolysis of the keto esters **51a** yields the ketocalixarenes (**51b**) which were tested for their ultraviolet light absorbing characteristics.

5.3.2   Oxidation of the Aryl Rings

Oxidation of calix[4]arene with Fremy's salt produces a burnt-orange colored solid with an ill-defined NMR spectrum, but oxidation with chlorine dioxide yields 83% of a bright yellow solid.[45] The ^1H NMR and mass spectra of this material are in agreement with a calix[4]quinone structure (**52**), but attempts to purify it have not been successful.

[43] A. A. Moshfegh, R. Badri, M. Hojjatie, M. Kaviani, B. Naderi, A. H. Nazmi, M. Ramezanian, B. Roozpeikar, and G. H. Hakimelahi, *Helv. Chim. Acta.*, **1982**, *65*, 1221; A. A. Moshfegh, B. Mazandarani, A. Nahid, and G. H. Hakimelahi, *ibid.*, **1982**, *65*, 1229; A. A. Moshfegh, E. Beladi, L. Radnia, A. S. Hosseini, S. Tofigh, and G. H. Hakimelahi, *ibid.*, **1982**, *65*, 1264.
[44] A. Ninagawa, K. Cho, and H. Matsuda, *Makromol. Chem.*, **1985**, *186*, 1379.
[45] L. Rosyk, Ph. D. Thesis, Washington University, St Louis, **1986**, p. 29.

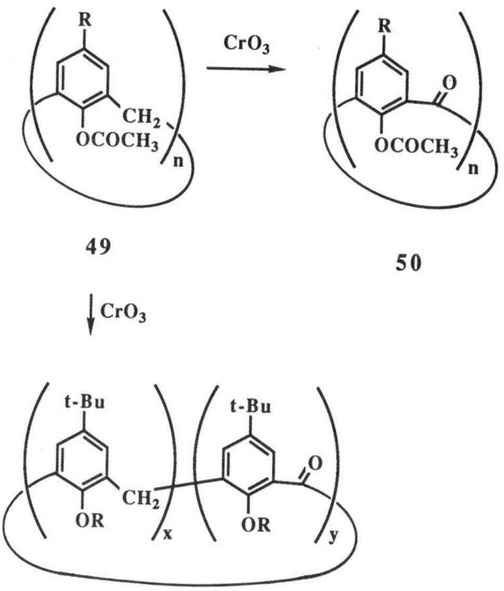

49   50

51a   R = OCOCH$_3$

51b   R = H

Figure 5.15   *Oxidation of methylene bridges of calixarenes*[43, 44]

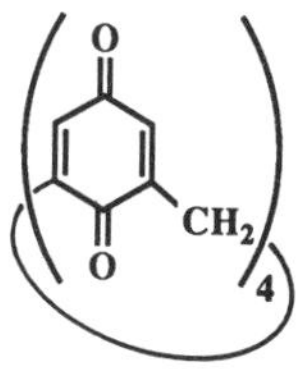

52

5.4   Concluding Remarks

The phenol-derived calixarenes carry hydroxyl groups at their 'lower rim', and these provide convenient handles for attaching other functional groups, including carboxyls which confer some water solubility on the compounds. Although direct 'upper rim' functionalization of the products from the one-step procedure is precluded in most cases, the easy removal of *tert*-butyl groups by a reverse Friedel–Crafts reaction makes the *para*-position available for facile substitution. Four procedures for effecting this have been used, *viz.* electrophilic substitution reactions, the *para*-Claisen rearrangement route, the *p*-quinonemethide route, and the *p*-chloromethylation route. These methods have made available a variety of functionalized phenol-derived calixarenes, including the water soluble *p*-sulfonatocalixarenes.

The resorcinol-derived calixarenes carry hydroxyl groups near their 'upper rim', and these provide sites for building bridges that hold the system in a conformationally immobile form. By means of electrophilic substitution reactions functional groups have been introduced into the aromatic rings, allowing the construction of compounds of unusual geometries. It is the full array of phenol-derived calixarenes and resorcinol-derived calixarenes, functionalized and unfunctionalized, that provides the basis for the discussions in the next two chapters which focus on the complexation and catalytic properties of calixarenes.

CHAPTER 6

Filling the Baskets: Complex Formation With Calixarenes

'Love, lay thy phobias to rest,
Inhibit thy taboo!
We twain shall share, forever blest,
A complex built for two

Keith Preston (1884–1927), *Love Song, Freudian*

The ability of calixarenes to act as baskets is one of their most intriguing properties, accounting for much of the interest that they have received in recent years. The discussions of the methods for making, shaping, and embroidering the calixarene baskets have been but prologue to the critical matter of testing their carrying capacities. This chapter deals with the various facets of calixarene complexation by first considering solid state systems and then proceeding to solution state systems involving aqueous as well as nonaqueous solvents.

6.1   Solid State Complexes

Many calixarenes form complexes in the solid state, this property having been observed even before the structures of the compounds had been established. Among the simple phenol-derived calixarenes, for example, *p-tert*-butylcalix[4]arene forms complexes with chloroform,[1] benzene,[2] toluene,[3] xylene,[2] and anisole;[2] *p-tert*-butylcalix[5]arene forms complexes with isopropyl alcohol[4] and acetone;[2] *p-tert*-butylcalix[6]arene forms a complex containing chloroform and methanol;[1] *p-tert*-butylcalix[7]arene forms a complex containing methanol;[5] and *p-tert*-butylcalix[8]arene forms a complex with chloroform.[1] The tenacity with which the guest molecule is held by the calixarene varies widely within the series, however. Whereas the cyclic tetramer and hexamer hold their guests very tightly, retaining residual amounts of the crystallization solvent even after long heating at high

[1] C. D. Gutsche, B. Dhawan, K. H. No, and R. Muthukrishnan, *J. Am. Chem. Soc.*, **1981**, *103*, 3782.
[2] M. Coruzzi, G. D. Andreetti, V. Bocchi, A. Pochini, and R. Ungaro, *J. Chem. Soc., Perkin Trans. 2*, **1982**, 1133.
[3] G. D. Andreetti, R. Ungaro, and A. Pochini, *J. Chem. Soc., Chem. Commun.*, **1979**, 1005.
[4] A. Ninagawa and H. Matsuda, *Makromol. Chem., Rapid Commun.*, **1982**, *3*, 65.
[5] Y. Nakamoto and S. Ishida, *Makromol. Chem., Rapid Commun.*, **1982**, *3*, 705.

temperature under vacuum, the cyclic octamer loses its guest chloroform upon standing a few minutes at room temperature and atmospheric pressure. The cyclic octamer crystallizes from chloroform as beautiful, glistening needles which quickly change to a white powder while the disappointed chemist watches ruefully. Numerous other members of the calixarene family also form complexes. To mention but two examples, Chasar's calix[4]arene[6] (see Figure 2.4) forms a dihydrate, and Asao's azulocalix[4]arene[7] (see Figure 2.11) retains benzene.

The structures of complexes of calixarenes, as those of the parent calixarenes, are most effectively revealed by *X*-ray crystallography. It was pointed out in an earlier chapter that particularly significant contributions in this area have been made by Andreetti, Pochini, and Ungaro in Parma. In the first of their many papers[3] on calixarenes they established the *endo*-calix character of the *p-tert*-butylcalix[4]arene–benzene complex, showing that the calixarene exists in the cone conformation and that the benzene guest molecule is ensconced exactly in the center of the cavity (see Figure 3.1). Similar demonstrations with nonaromatic guest molecules have more recently been provided by Atwood and coworkers[8] and by McKervey and coworkers.[9] The Atwood group showed that the pentaammonium salt of *p*-sulfonatocalix[4]arene (see **22** in Figure 5.8) forms an *endo*-calix complex with methyl sulfate, and the McKervey group prepared the complex of acetonitrile and the tetracarbonate of *p-tert*-butylcalix[4]arene (see **12**, $R^1 = CO_2Et$, $R_2 = tert$-butyl in Figure 5.5). The acetonitrile molecule, inside the cavity, was shown to be oriented with the nitrogen pointing outward and the methyl group *ca.* 3.6 Å from carbons C-25,26,27,28 at the bottom of the calix. In contrast, Rizzoli *et al.*,[10] found that the tetraacetate of *p-tert*-butylcalix[4]arene assumes a 1,3-alternate conformation upon forming a complex with acetic acid dimer. In the terminology of Weber and Josel[11] all of these examples are classed as 'crypt-ato-cavitate clathrato' complexes or, as Ungaro *et al.*[12] argue, as true molecular complexes because of the nature of the complexing forces and the fact that the calix is retained in the absence of the guest, even in solution (although in solution it is conformationally mobile). Subsequent work has revealed that other calixarenes may form different kinds of complexes, and several varieties of *exo*-calix complexes have now been characterized. For example, *p*-(1,1,3,3-tetramethylbutyl)calix[4]arene forms an intercalato-clathrate in which the toluene is captured *between* the molecules of calixarene rather than within the calix.[13] This is the result of flexible *p*-substituents in the calixarene, two of which can bend inward to fill the cavity, as shown in Figure 6.1. Intramolecu-

---

[6] D. W. Chasar, *J. Org. Chem.*, **1985**, *50*, 545.

[7] T. Asao, S. Ito, and N. Morita, *Tetrahedron Lett.*, **1988**, *29*, 2839.

[8] S. G. Bott, A. W. Coleman, and J. L. Atwood, *J. Am. Chem. Soc.*, **1988**, *110*, 610.

[9] M. A. McKervey, E. M. Seward, G. Ferguson, and B. L. Ruhl, *J. Org. Chem.*, **1986**, *51*, 3581.

[10] C. Rizzoli, G. D. Andreetti, R. Ungaro, and A. Pochini, *J. Mol. Struct.*, **1982**, *82*, 133.

[11] E. Weber and H.-P. Josel, *J. Inclusion Phenom.*, **1983**, *1*, 79.

[12] R. Ungaro, A. Pochini, G. D. Andreetti, and P. Domiano, *J. Chem. Soc., Perkin Trans. 2*, **1985**, 197.

[13] G. D. Andreetti, A. Pochini, and R. Ungaro, *J. Chem. Soc., Perkin Trans. 2*, **1983**, 1773.

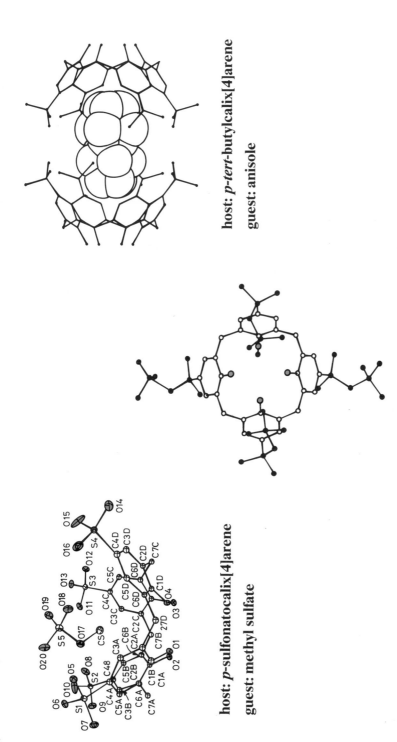

host: *p-tert*-butylcalix[4]arene
guest: anisole

host: *p*-sulfonatocalix[4]arene
guest: methyl sulfate

Figure 6.1 X-Ray crystallographic structures of calixarenes and calixarene complexes[8,12,13]

lar complexation thus prevents an external guest molecule, such as toluene, from entering the cavity. When this same calixarene is crystallized from acetone it produces a guest-free compound which, nevertheless, still retains the cone conformation. Calix[4]arene, carrying hydrogens in the *para*-positions, forms a pair of complexes with acetone.[14] One is a 1:3 complex that is characterized as the 'hexagonal crystal form' and classed as an intercalate-clathrate in which the cone is considerably flattened. The authors state that the building block of the crystal is a calix[4]arene–acetone couple in which a methyl group of the acetone points into the calix. The other is a 1:1 complex, characterized as the 'orthorhombic crystal form', that is classed as a tubulato-clathrate in which columns are formed that are stated[10] to possess 6_3 symmetry and in which 'a polar interior part is formed by the facing oxygen atoms of the calixarene molecules'. Still another mode of complexation has been discerned in the interaction of *p-tert*-butylcalix[4]arene and anisole which leads to a 2:1 complex.[12] An *X*-ray crystallographic determination shows it to be a double *endo*-calix complex in which a single anisole molecule is shared by a pair of calixarenes which are oriented rim-to-rim to form a closed cavity encapsulating the anisole, as pictured in Figure 6.1. The crown-bridged *p-tert*-butylcalix[4]arene (*cf.* compound **8b** in Section 4.3.1.2) forms a 1:1 complex with pyridine in which the pyridine molecule is oriented with its nitrogen proximate to a *tert*-butyl group.[15]

An analytical technique that has been infrequently applied to molecular complexes but that has the potential for supporting and complementing the information gained from *X*-ray crystallography is that of ^{13}C cross-polarization magic angle spinning NMR. Komoto *et al.*[16] have used this technique to investigate the toluene complex of *p-tert*-butylcalix[4]arene in which they noted a 6.2 ppm upfield shift in the resonance of the toluene methyl group. Employing the Johnson and Bovey calculations,[17] they concluded that the toluene is positioned in the cavity of the calixarene with the methyl carbon of the guest 3.20 Å above the center of the phenyl rings of the calixarene, in good agreement with the structure derived by *X*-ray analysis.

Calix[5]arene forms a 1:2 complex with acetone that is shown by *X*-ray crystallography[2] to contain one acetone molecule within the calix and another external to it. Thus, it is both an *endo*-calix and an *exo*-calix complex, both a cryptato-cavitate complex and an intercalato-clathrate. The octamethyl ether of *p-tert*-butylcalix[8]arene forms a similarly loose complex with chloroform, but Atwood and coworkers[18] have successfully obtained its *X*-ray structure which shows two molecules of chloroform to be held in a quasi *endo*-calix position, as illustrated in Figure 6.2. The conformation of the

[14] R. Ungaro, A. Pochini, G. D. Andreetti, and V. Sangermano, *J. Chem. Soc., Perkin Trans. 2,* **1984**, 1979.

[15] G. D. Andreetti, O. Ori, F. Ugozzoli, C. Alfieri, A. Pochini, and R. Ungaro, *J. Inclusion Phenom.*, **1988**, *6*, 523.

[16] T. Komoto, I. Ando, Y. Nakamoto, and S.-i. Ishida, *J. Chem. Soc., Chem. Commun.*, **1988**, 135.

[17] C. D. Johnson, Jr. and F. A. Bovey, *J. Chem. Phys.*, **1958**, *29*, 1012.

[18] A. W. Coleman, S. G. Bott, and J. L. Atwood, *J. Inclusion Phenom.*, **1986**, *4*, 247.

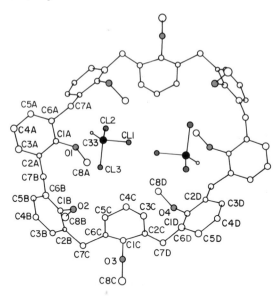

Figure 6.2 X-*Ray crystallographic structure of the CDCl₃ complex of the octa-methyl ether of* p-tert-*butylcalix[8]arene*[18]

cyclic octamer in this complex is one in which six of the methoxyl groups are oriented toward the inside of the cavity and two on the outside, in contrast to the '4 in, 4 out' arrangement of the octamethoxyethyl ether of p-(1,1,3,3-tetramethylbutyl)calix[8]arene,[19] the '6 in, 2 out' arrangement for the octa-acetate of p-*tert*-butylcalix[8]arene,[20] and the 'all in' arrangement of p-*tert*-butylcalix[8]arene itself.[21]

The work of the Parma group has provided interesting insight into some of the factors responsible for solid state complexation, or lack thereof, showing that it is dependent on the size of the calixarene,[2,14] the particular substituents in the *para*-positions, and the conformational rigidity of the calixarene.[12] *tert*-Butyl groups prove to be particularly favorable for the formation of *endo*-calix complexes, ostensibly because (a) they are not flexible enough to form intramolecular complexes comparable to those of the p-(1,1,3,3,-tetramethylbutyl)calix[4]arenes (see above) and (b) they permit CH_3/π interaction[22] to occur between the protons of the *tert*-butyl groups and the aromatic ring of the guest molecule, this feature being particularly well illustrated by the aforementioned pyridine complex.[15] It is fascinating, though often frustrating, to see how sensitive the formation of clathrates is to subtle

[19] R. Ungaro, A. Pochini, G. D. Andreetti, and F. Ugozzoli, *J. Inclusion Phenom.*, **1985**, *3*, 409.
[20] G. D. Andreetti, R. Ungaro, and A. Pochini, *J. Chem. Soc., Chem. Commun.*, **1981**, 533.
[21] C. D. Gutsche, A. E. Gutsche, and A. I. Karaulov, *J. Inclusion Phenom.*, **1985**, *3*, 447.
[22] J. Uzawa, S. Zushi, Y. Kodama, Y. Fukuda, K. Nishihata, K. Umemura, M. Nishio, and M. Hirota, *Bull. Chem. Soc. Jpn.*, **1980**, *53*, 3623.

structural changes in the host and guest molecules. Similar phenomena have been observed by other workers in other systems.[23,24]

A number of metal-containing calixarenes, themselves often referred to as 'complexes' but treated in Chapter 4 (see Section 4.3.1.3) as 'lower rim' bridged compounds, form noncovalent complexes of the type under discussion. The titanium compound shown in Figure 4.21 that was synthesized by Atwood and coworkers[25] crystallizes with four molecules of toluene which occupy *exo*-calix lattice positions. The *p-tert*-butylcalix[4]arene tetramethyl ether–toluene–Na$^+$ complex shown in Figure 4.23 contains a toluene molecule inside the cavity (*i.e.* an *endo*-calix complex) positioned so that its methyl group is directly above the sodium at a separation of 4.35 Å and its aromatic carbons are in contact with the *tert*-butyl groups of the calixarene with an average separation of 3.77 Å. The titanium complex of *p-tert*-butylcalix[6]arene synthesized by Andreetti and coworkers[26] (see Figure 4.19) forms an *endo*-calix complex with toluene.

The joining of spherand chemistry and calixarene chemistry was accomplished by the European consortium[27] of Reinhoudt *et al.* in Holland and Ungaro *et al.* in Italy who synthesized a calixspherand (see Figure 4.16). The product contains a molecule of NaBr, indicating its strong propensity to form complexes with cations. An *X*-ray structure (on the picrate salt, obtained by ion exchange) showed the Na$^+$ to be located in the middle of the cavity created at the 'lower rim' of the calix by the hemispherand bridge, as illustrated in Figure 6.3. The calixarene portion of the calixspherand is forced into a partial cone conformation in the complex, resulting in O—Na$^+$ distances varying from 2.38—2.60 Å and creating a cavity of *ca.* 2.2 Å in diameter which is larger than the 1.96 Å diameter of Na$^+$. To remove the Na$^+$ from the complex it is necessary to heat the compound in a 1:4 methanol water mixture at 120 °C in a sealed ampule, the driving force for the decomplexation being the crystallization of the calixspherand from the medium.

As part of the comprehensive program dealing with host–guest chemistry for which he shared the 1987 Nobel Prize, Donald Cram studied the solid state complexation characteristics of cavitands and carcerands. The bridged calix[4]resorcinarenes (see **14** and **15** in Figure 4.25), which Cram calls cavitands, form solid state complexes with certain small molecules. In the case of compound **14** (X = CH$_2$) in Figure 4.25 the complex with chloroform[28] is stable enough to resist decomposition at 100 °C and 10^{-5} mm Hg. The *X*-ray structure of the complex of **5a** (*cf.* p. 169) and carbon disulfide[29]

[23] D. D. MacNicol, J. J. McKendrick, and D. R. Wilson, *Chem. Soc. Rev.*, **1978**, *7*, 65.

[24] R. Hilgenfeld and W. Saenger, *Angew. Chem., Suppl.*, **1982**, 1690.

[25] S. G. Bott, A. W. Coleman, and J. L. Atwood, *J. Chem. Commun.*, **1986**, 610; *idem., J. Am. Chem. Soc.*, **1986**, *108*, 1709.

[26] G. D. Andreetti, G. Calestani, F. Ugozzoli, A. Arduini, E. Ghidini, A. Pochini, and R. Ungaro, *J. Inclusion Phenom.*, **1987**, *5*, 123.

[27] D. N. Reinhoudt, P. J. Dijkstra, P. J. A. in't Veld, K. E. Bugge, S. Harkema, R. Ungaro, and E. Ghidini, *J. Am. Chem. Soc.*, **1987**, *109*, 4761.

[28] J. R. Moran, S. Karbach, and D. J. Cram, *J. Am. Chem. Soc.*, **1982**, *104*, 5826.

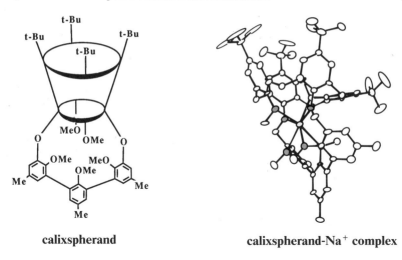

calixspherand   calixspherand-Na$^+$ complex

Figure 6.3   *Structure of a calixspherand and its complex with Na$^+$* [27]

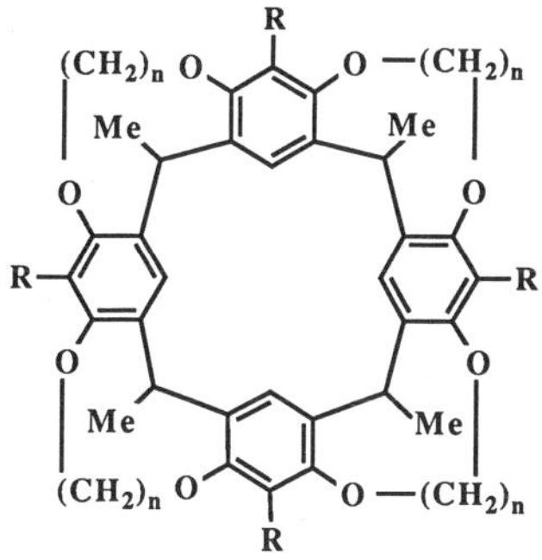

a	R = H, n = 1
b	R = CH$_3$, n = 1
c	R = Br, n = 1
d	R = I, n = 1
e	R = CH$_3$, n = 2
f	R = Br, n = 2
g	R = CH$_3$, n = 3
h	R = Br, n = 3

1

Figure 6.4   *Resorcinol-derived bridged calixarenes (cavitands)* [30]

The *X*-ray structure of the complex of **5a** (*cf.* p. 169) and carbon disulfide[29] clearly shows its *endo*-calix structure. In a subsequent and more detailed study the Los Angeles group[30] prepared a series of compounds of the general structure **1** shown in Figure 6.4 in which they varied the circumference of the cavity by changing *n* and varied the depth of the cavity by changing **R**. Using *X*-ray crystallography, the exact geometries of the complexes formed from these various host and guest molecules were determined. Included as guest molecules were acetonitrile, chloroform, methylene chloride, benzene, toluene, and cyclohexane. The **1a**−CH$_2$Cl$_2$ complex, for example, showed the

[29] D. J. Cram, K. D. Stewart, I. Goldberg, and K. N. Trueblood, *J. Am. Chem. Soc.*, **1985**, *107*, 2574; I. Goldberg, *J. Inclusion Phenom.*, **1986**, *4*, 191.

[30] D. J. Cram, S. Karbach, H.-E. Kim, C. B. Knobler, E. F. Maverick, J. L. Ericson, and R. C. Helgeson, *J. Am. Chem. Soc.*, **1988**, *110*, 2229.

CH_2Cl moiety of the guest to lie mostly in the open end of the cavity, with the other Cl protruding from the rim. In the **1b**–CH_3CN complex, where the host is deeper, the CH_3C penetrates far enough into the cavity to bring the tops of the OCH_2O hydrogens almost level with the nitrogen atom of the guest. In the complexes **1c**–$CHCl_3$ and **1f**–CH_2Cl_2, one of the chlorines of the host lies deep in the cavity with its C—Cl dipole aligned to complement the Br—C dipole of the host. A particularly interesting result was observed in the ternary complex **1b**–benzene–cyclohexane in which two hydrogens and a carbon (CCH_2C) of a *boat* cyclohexane penetrate the top of the cavity, the host apparently forcing the cyclohexane into a boat conformation. The authors conclude from this study that 'cavitands of varying dimensions containing a variety of substituents can be designed with the aid of CPK molecular models and then readily prepared', a conclusion that seems to be well substantiated in this instance and that undoubtedly will be extended to a variety of other systems in the future.

Probably the most unusual of all of the host–guest complexes that have been synthesized are those involving the carcerands, which are structures consisting of a pair of cavitands joined 'upper rim' to 'upper rim', as pictured in Figure 5.14 and shown schematically in Figure 6.5. When the joining of the two cavitands takes place it creates a vault large enough to contain some of the molecules present in the reaction milieu. Thus, treatment[31] of a dioxane–DMF solution containing cavitand-A (the tetrachloro compound) and cavitand-B (the tetrathiol) treated with pulverized Cs_2CO_3 in an argon atmosphere gives a 29% yield of a gray-white powder that is insoluble in everything, including hot naphthalene, anisole, nitrobenzene, pyridine, or xylene. Elemental analysis shows that it contains C, H, N, S, Cl, Cs, and Ar. A molecular formula that fits the percentage to which these elements are present is $C_{83.84}H_{77.99}O_{21.18}S_4N_{0.39}Cl_{0.35}Cs_{0.35}Ar_{0.0065}$, yielding a possible structure of $\{(C_{80}H_{72}O_{16}S_4):0.39 \ DMF:0.67 \ dioxane:0.35 \ CsCl:0.0065 \ Ar:O_{4.12}\}$.

CPK Model of a carcerand

[31] D. J. Cram, S. Karbach, Y. H. Kim, L. Baczynskyj, K. Marti, R. M. Sampson, and G. W. Kalleymeyn, *J. Am. Chem. Soc.*, **1988**, *110*, 2554.

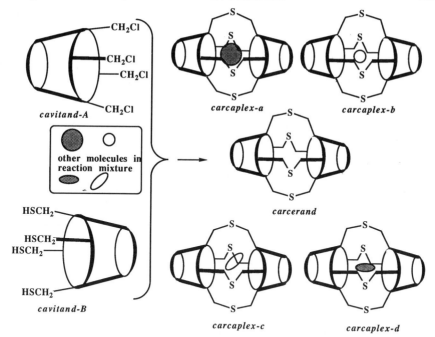

Figure 6.5 *Conversion of a cavitand into a carcerand and carcaplexes[31]*

Thus, the solvent molecules, the cesium ion, and the argon all are captured inside the cavity during the joining process, securely incarcerated within this vault to give what Cram calls a 'carcaplex'. Only 4.12 oxygens remain unaccounted for, and the authors speculate that these may come from adventitious silica acquired from the sintered glass funnel used in the isolation. The authors add the *caveat* that this treatment of the analytical data is reasonable but not unique and that alternative interpretations in which water molecules are substituted for some of the dioxane molecules also give good agreement between the calculated and observed values.

The synthesis of the carcerand from cavitands of well-established structure, taken in tandem with the unusual complexation character of the carcerand, provides good presumptive evidence for its structure. Nevertheless, direct evidence is always desired, although acquiring it can be a difficult task in cases where the product is so intractable. Fortunately, both mass spectral and NMR techniques are available to deal with such recalcitrant molecules. Thus, fast atom bombardment mass spectrometry of the product shows strong signals for molecules corresponding to the parent ion, the (parent ion + DMF), the (parent ion + dioxane), the (parent ion + Cs), and the (parent ion + H_2O). A particularly interesting feature of the mass spectrum is the observation both of a carcerand–Cs species and a carcerand–CsCl species, the first containing only Cs^+ inside the cavity and the second containing cesium as well as a counterion Cl^- inside the cavity. Although solution state NMR is inapplicable to this insoluble molecule, solid state NMR using crossed-polarization magic angle spinning (CPMAS) provides a ^{13}C NMR spectrum commensurate with the carcerand structure.

The carcerand is a 'molecular cell' containing pores through which substances might pass or not, depending on their size. That water can do so was suggested by signals in the mass spectrum. The ability of water to enter the cavity *after* it is formed was verified by treating the carcerand with D_2O and observing that the signals in the mass spectrum arising from the water-containing carcaplexes then show the presence of D_2O. The slowness with which water molecules enter the cavity, however, suggests that there is a fairly high energy barrier for this process. Oxygen, also, appears to be able to pass into the cavity, and this provides an explanation for the combustion-fragmentation behavior that is noted on heating the carcerand. It is postulated that oxygen inside the cavity oxidizes the sulfide to sulfoxide and/or sulfone groups, causing the vault to crack open, lose its contents, and become more susceptible to conventional combustion pathways. Argon, larger than oxygen, cannot traverse the pores of the carcerand and is trapped inside during the synthesis approximately once in every 150 shell closings. The diameter of argon is *ca.* 3.10 Å, which is greater than the diameter of 2.59 Å estimated for the pores of the host. Cesium, with a diameter of 3.4 Å, also is too large to traverse the pores and once entrapped becomes a permanent prisoner until the carcerand is decomposed.* The especially large amount of incarcerated cesium in the mixture of carcaplexes prepared as described above is ascribed to the geometry of the S_N2 spanner-forming reactions. These require the cesium, coordinated with the sulfur atom, to be inside the cavity and the chlorine attached to the carbon that is undergoing attack by the nucleophilic sulfur to be on the outside. Thus, when the fourth bond juncture is made the cesium is trapped.

6.2   Solution State Complexes in Nonaqueous Solvents

6.2.1   Complexes with Metal Cations

The calixarenes that were initially available, either by one-step or multi-step procedures, are compounds that are almost totally insoluble in water and only sparingly soluble in most organic solvents. The first success in demonstrating their complexing abilities was achieved by Reed Izatt and his colleagues at Brigham Young University in Provo, Utah. Izatt, who has carried out extensive investigations on the complexation behavior of crown ethers and various related types of compounds, perceived a structural resemblance between the crown ethers, cyclodextrins, and calixarenes (see space-filling models, p. 203) and proceeded to test the latter for their ability to transport cations across a liquid membrane. They employed an apparatus in which an aqueous source phase containing the host molecule (the carrier) and the

---

*I had to sink my yacht to make the guests go home' — Ernest Hemingway, *Notebooks*.

Table 6.1 *Cation transport from basic solution by calixarenes (data given in flux in* moles $s^{-1}/m^2 \times 10^8)^{32,33}$

Source phase	Cation diameter/Å	p-tert-Butylcalixarene			p-tert-Pentylcalixarene	
		[4]	[6]	[8]	[4]	[8]
LiOH	1.52	—	10	2	—	—
NaOH	2.04	2	22	9	3	10
KOH	2.76	<0.7	13	10	1	23
RbOH	3.04	6	71	340	12	111
CsOH	3.40	260	810	996	414	92
Ba(OH)$_2$	2.70	1.6	3.2	—		

cations is separated by an organic phase (*e.g.* chloroform) from an aqueous receiving phase. What they discovered[32,33] is that although the calixarenes are ineffective cation carriers in neutral solution, they possess significant transport ability for Group I cations in strongly basic solution, as shown by the data in Table 6.1. This is in sharp contrast to 18-crown-6 compounds which are more effective in neutral than in basic solution, *e.g.* KNO$_3$ *vs.* KOH. Group II cations, including Ca^{2+}, Ba^{2+}, and Sr^{2+} are not effectively transported by these calixarenes. Control experiments with *p-tert-*butylphenol, which shows little or no transport ability itself, support the idea that the macrocyclic ring plays a critical role, although it is not yet clear exactly what that role is. The diameters of the annuli of the calixarene mono-anions are *ca.* 1.0 Å for the cyclic tetramer, 2.4 Å for the cyclic hexamer, and 4.8 Å for the cyclic octamer. Thus, for complexes in which the cation and oxygens are coplanar, the cyclic tetramer has too small an opening even for Li$^+$, whereas the cyclic octamer has too large an opening to fit snugly even around Cs$^+$. The cyclic hexamer seems to be the system best constituted to behave in crown ether-like fashion and, indeed, even the synthesis of *p-tert-*butylcalix[6]arene may be influenced by a 'template effect' as discussed in Chapter 2. The dramatic difference between the ability of all three of the even-numbered calixarenes to transport Cs$^+$ as compared with the other cations, however, must depend on something other than this simple complementarity. On the basis of a comparison of the ion transport capacity of a series of 'upper rim' bridged calixarenes (see **12** in Figure 4.24) Böhmer and coworkers[34] have suggested that at least in the case of cesium the transport is by an *endo*-calix complex. Figure 6.6 shows the variation in cesium ion transport rate as a function of the length of the 'upper rim' spanner, maxima being observed for *n* = 8 and *n* = 14. Böhmer argues that these results cannot be understood in terms of interactions of Cs$^+$ with the oxygen atoms and are better interpreted as arising from perturbations of the calix induced by the

[32] R. M. Izatt, J. D. Lamb, R. T. Hawkins, P. R. Brown, S. R. Izatt, and J. J. Christensen, *J. Am. Chem. Soc.*, **1983**, *105*, 1782.
[33] S. R. Izatt, R. T. Hawkins, J. J. Christensen, and R. M. Izatt, *J. Am. Chem. Soc.*, **1985**, *107*, 63.
[34] H. Goldman, W. Vogt, E. Paulus, and V. Böhmer, *J. Am. Chem. Soc.*, **1988**, *110*, 6811.

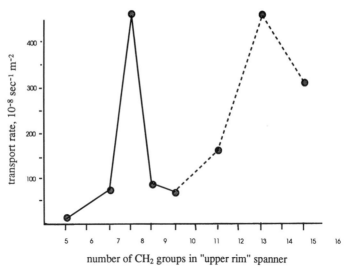

number of CH$_2$ groups in "upper rim" spanner

Figure 6.6   *Cs$^+$ transport rate as a function of the length of the spanner in 'upper rim' bridged calix[4]arenes[34]*

spanner, an *endo*-calix complex providing a particularly good environment for a Cs$^+$ after it has lost its hydration shell.

Extending their initial studies to include multi-carbon systems comprising equimolar mixtures of two, three, and four cations from NaOH, KOH, RbOH, and CsOH, Izatt and coworkers[33] found that selective transport of Cs$^+$ occurs in all cases. As in the single ion system, the greatest selectivity for Cs$^+$ is shown by the calix[4]arenes, but the largest flux is found in the calix[8]arenes. That the cation flux depends, at least in part, on the relative concentrations of the cations in the source phase is illustrated by the fact that at low Cs$^+$:Rb$^+$ ratios *p-tert*-butycalix[6]arene transports Rb$^+$ more rapidly than Cs$^+$. As this ratio increases, however, the transport of Cs$^+$ becomes favored. Also demonstrated is a synergistic effect in which the presence of Cs$^+$ in a two- or three-cation mixture with Na$^+$ significantly increases the Na$^+$ flux over that when Na$^+$ is present alone.

The studies of the Provo group point out the useful features of the calixarenes as ion carriers, *viz.* their low water solubility, their ability to form neutral complexes with cations through the loss of protons, and their potential for allowing the coupling of cation transport with the reverse flux of protons. Most of the subsequent work from other laboratories has focused not on the calixarenes themselves but on various derivatives. One of the earliest of these was the observation by Bocchi *et al.*[35] that while the tetraacetate of *p-tert*-butylcalix[4]arene fails to form complexes with guanidinium ion or Cs$^+$, the octa-(3,6-dioxaheptyl)ether of *p-tert*-butylcalix[8]arene forms strong complexes with these cations. The Parma group later showed[36] that the hexa-

[35] V. Bocchi, D. Foina, A. Pochini, and R. Ungaro, *Tetrahedron*, **1982**, *38*, 373.
[36] R. Ungaro, A. Pochini, G. D. Andreetti, and P. Domiano, *J. Inclusion Phenom.*, **1985**, *3*, 35.

Table 6.2 *Extraction of alkali metal picrates in CH_2Cl_2 at 20 °C by carboalkoxymethyl ethers of calixarenes (**2a–2l**). Values without parentheses are from McKervey et al.,[37] values in parentheses are from Chang and Cho[42,43]*

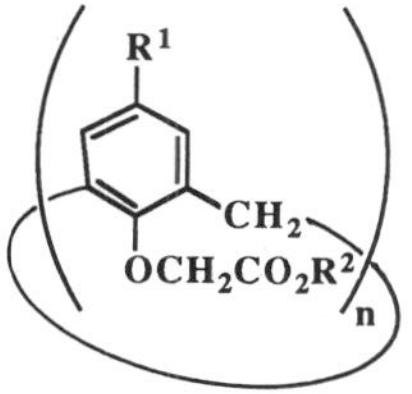

2

a R¹ = tert-butyl, R² = Et, n = 4
b R¹ = tert-butyl, R² = Me, n = 4
c R¹ = H, R² = Et, n = 4
d R¹ = H, R² = Me, n = 4
e R¹ = tert-butyl, R² = Et, n = 6
f R¹ = tert-butyl, R² = Me, n = 6

g R¹ = H, R² = Et, n = 6
h R¹ = H, R² = Me, n = 6
i R¹ = tert-butyl, R² = Et, n = 8
j R¹ = tert-butyl, R² = Me, n = 8
k R¹ = H, R² = Et, n = 8
l R¹ = H, R² = Me, n = 8

Calix[4]arenes

	2a	2b	2c	2d
Li⁺	15.0(48.9)	6.7	1.8	1.1
Na⁺	94.6(87.7)	85.7	60.4	34.2
K⁺	49.1(51.2)	22.3	12.9	4.8
Rb⁺	23.6(41.0)	9.8	4.1	1.9
Cs⁺	48.9(52.8)	25.5	10.6	10.8

Calix[6]arenes

	2e	2f	2g	2h
Li⁺	11.4(6.7)	1.7	4.7	2.6
Na⁺	50.1(15.6)	10.3	10.4	6.7
K⁺	85.9(66.2)	29.1	51.3	25.2
Rb⁺	88.7(60.5)	41.2	94.1	77.7
Cs⁺	100.0(88.9)	54.8	94.6	94.6

Calix[8]arenes

	2i	2j	2k	2l
Li⁺	1.1(<1)	0.9	0.8	0.4
Na⁺	6.0(4.5)	8.3	7.5	4.1
K⁺	26.0(21.5)	25.5	20.2	12.1
Rb⁺	30.2(16.4)	29.8	28.9	17.5
Cs⁺	24.5(17.0)	20.1	30.1	27.0

(3-oxabutyl) ether of *p-tert*-butylcalix[6]arene also forms complexes with these ions. An extensive survey has been carried out by McKervey and his coworkers[37] who prepared the carboalkoxymethyl ethers of *p-tert*-butylcalix[4]arene,   *p-tert*-butylcalix[6]arene,   *p-tert*-butylcalix[8]arene, calix[4]arene, calix[6]arene, and calix[8]arene, and measured their abilities to extract cations from the aqueous phase into the nonaqueous phase. The results, shown in Table 6.2, reveal a wide range of phase-transfer efficiency among these twelve ethers and provide the basis for the following géneralizations: (a) the calix[4]arene compounds show the greatest selectivity for Na^+; (b) phase-transfer of Li^+ is inefficient with all of the compounds; (c) the calix[6]arene compounds show less affinity for Na^+ than for K^+, with plateau selectivity for Rb^+ and Cs^+; (d) the calix[8]arene compounds are the least efficient of the cyclic oligomers, showing low levels of transport and low discrimination for all five cations; (e) the calix[6]arene compounds are significantly more effective than 18-crown-6 for Na^+ and K^+ and much more so for Cs^+; (f) the *tert*-butyl group appears to increase to some extent the selectivity of the calix[4]arene compounds for Na^+. That there is truly a macrocyclic effect in operation is demonstrated by the very low level of ion transport effected by a linear tetramer analog of the calix[4]arene compound. It is interesting to note also that the diether analog of **2a** is ineffective.

An even better cation transport compound is the keto ether **3** which gives the following extraction values:[38] Li^+ (31.4%), Na^+ (99.2%), K^+ (84.1%), Rb^+ (53.7%), Cs^+ (83.8%), and NH_4^+ (23.0%). Comparing these values with

3

those for **2a**, it is seen that in every case **3** is more effective than **2** (left-hand vertical column in Table 6.2; the value for **2** with NH_4^+ is less than 5%). The keto ether is a remarkably good ion extracting compound that undoubtedly will inspire further work along these lines. Another recently introduced compound is the diethylamide of the tetracarboxymethyl ether of *p-tert*-butylcalix[4]arene for which Calestani *et al.*[39] report alkali metal picrate extraction constants of 1.9×10^9 for Na^+, 2.8×10^7 for K^+, and 1.3×10^7 for

[37] M. A. McKervey, E. M. Seward, G. Ferguson, B. Ruhl, and S. J. Harris, *J. Chem. Soc., Chem. Commun.*, **1985**, 388.
[38] G. Ferguson, B. Kaitner, M. A. McKervey, and E. M. Seward, *J. Chem. Soc., Chem. Commun.*, **1987**, 584.
[39] G. Calestani, F. Ugozzoli, A. Arduini, E. Ghidini, and R. Ungaro, *J. Chem. Soc., Chem. Commun.*, **1987**, 344.

Li$^+$, constants of a magnitude that persuaded the authors to liken the binding properties of these calixarenes to cryptands and spherands. As a selective reagent for Na$^+$, Arduini *et al.*[40] recommend the *tert*-butyl ester of the tetra-carboxymethyl ether of *p-tert*-butylcalix[4]arene [see **12** in Figure 5.5, R^1 = CH$_2$CO$_2$C(CH$_3$)$_3$ and R^2 = *tert*-butyl], and this same group has shown[41] that the crown ether-calixarene dicarboxylic acid (see **11** in Figure 5.4) forms a complex with Ca^{2+} in which the metal ion is inside the polyether cavity.

Concomitant with McKervey's work, Chang and Cho[42] in Seoul, Korea carried out similar studies. Their preliminary results, which were published a few months earlier than those from Ireland, showed the extraction values for **2e** and **2i** indicated in parentheses in Table 6.2. In a subsequent paper[43] the values for **2a** were also published. Although there is lack of quantitative agreement between the results from the Irish and Korean laboratories, the trends and comparative values are similar in most, though not all, instances. The Korean group also tested the analogous methyl and ethoxyethyl ethers of *p-tert*-butylcalix[4]arene, *p-tert*-butylcalix[6]arene, and *p-tert*-butyl-calix[8]arene and found them to be essentially devoid of extraction capabilities. Compounds **2e** and **2i** were tested against Ca^{2+} and Ba^{2+} and found to have extraction values of 5.3—6.4% and 9.2—17.9%, respectively. By comparing the transport abilities of the parent calixarenes, their carbethoxy-methyl ethers, and their methyl ethers using the analytical technique described by Izatt *et al.*, this study brings together the work of the American, Irish, and Korean groups. The results, as seen in Table 6.3, reaffirm the enhanced ability of the carbethoxymethyl ethers to interact effectively with Group I cations, the transport capacity of **4** (R = CH$_2$CO$_2$Et, n = 4,6,8) being very much greater than that of **4** (R = H, n = 4,6,8). The marked contrast between the ability of the calixarene methyl ethers (**4**, R = Me) and Cram's spherands[44] to form metal cation complexes can be attributed to the far greater preorganization (*i.e.* rigidity) of the spherands as compared with the calix[6]arenes and calix[8]arenes.

Chang and coworkers[45] have carried the theme of calixarene–cation interaction a further step by testing the ion carrying capacities of the amides **4** (R = CH$_2$CONHC$_4$H$_9$), prepared as described in Chapter 5 from the corresponding esters **4** (R = CH$_2$CO$_2$Et). Although the amides are much less effective than the esters for the complexation of Group I cations, the data in Table 6.3 show that they are better for Group II cations, as anticipated. That the binding site in the complexation of the calixarene ethers **4** (R = CH$_3$) with metals is at or near the 'lower rim' of the calix was indicated by the large downfield shifts of the ArCH$_2$Ar and OCH$_2$ methylene resonances in the presence of an europium NMR shift reagent.

[40] A. Arduini, A. Pochini, S. Reverberi, and R. Ungaro, *Tetrahedron*, **1986**, *42*, 2089.
[41] R. Ungaro, A. Pochini, and G. D. Andreetti, *J. Inclusion Phenom.*, **1984**, *2*, 199.
[42] S.-K. Chang and I. Cho, *Chem. Lett.*, **1984**, 477.
[43] S.-K. Chang and I. Cho, *J. Chem. Soc., Perkin Trans. 1*, **1986**, 211.
[44] D. J. Cram, G. M. Lein, T. Kaneda, R. C. Helgeson, C. B. Knobler, E. Maverick, and K. N. Trueblood, *J. Am. Chem. Soc.*, **1981**, *103*, 6228.
[45] S.-K. Chang, S.-K. Kwon, and I. Cho, *Chem. Lett.*, **1987**, 947.

Table 6.3 *Transport rates (bold numbers, moles $\times 10^{-7}$ h^{-1}) and extraction efficiencies (italicized numbers in % of picrate salt extracted) for calixarenes and calixarene ethers[43, 45]*

4

R	n	Na$^+$	K$^+$	Cs$^+$	Mg^{2+}	Ca^{2+}	Sr^{2+}	Ba^{2+}
H	4	**2.0**	**2.3**	**1.2**				
H	6	**1.0**	**2.5**	**1.4**				
H	8	**8.6**	**3.6**	**4.0**				
CH$_2$CO$_2$Et	4	**140**	**71**	**21.9**				
CH$_2$CO$_2$Et	6	**106**	**473**	**556**	—	*5.3*	—	*8.2*
CH$_2$CO$_2$Et	8	**25.8**	**100**	**75.2**	*<1*	*6.4*	—	*17.9*
Me	4	**4.1**	**10.9**	**3.2**				
Me	6	**11.1**	**10.6**	**21.2**				
Me	8	**22.5**	**24.2**	**20.8**				
CH$_2$CONHBu	4	*2.7*	*<1*	*>3.1*	*<1*	*<5.8*	*4.6*	*4.8*
CH$_2$CONHBu	6	*3.1*	*<1*	*2.9*	*11.8*	*33.4*	*56.6*	*30.2*
CH$_2$CONHBu	8	*1.4*	*5.4*	*3.6*	*14.2*	*16.4*	*28.6*	*37.0*

The calixspherand discussed earlier in this chapter as a solid state complex with Na$^+$ (see Figure 6.3) has been shown to also engage cations in solution.[27] Extraction experiments with alkali picrates give the following binding energies ($-\Delta G°$ in kcal/mole): Na$^+$ 13.6 ± 0.2, K$^+$ 14.0 ± 0.2, Rb$^+$ 12.0 ± 0.2, Cs$^+$ 9.8 ± 0.2. The rates of complexation, however, fall in a different order, *viz.* Rb$^+$ and Cs$^+$ ≤ 6 h, K$^+$ ≥ 12 h, Na$^+$ ≥ 48 h, indicating that complete desolvation must occur prior to complexation.[46] A related compound in which a crown ether rather than a hemispherand moiety bridges the 1,3-oxygens of *p-tert*-butylcalix[4]arene (dimethyl ether of compound **8b** on p. 114) proves to be an effective ionophore for K$^+$, Rb$^+$, and Cs$^+$ with $-\Delta G°$ values of 11.3, 10.6, and 7.8 kcal/mole, respectively. A selectivity of $K^{K^+}/K^{Na^+} = 2000$ was observed.

6.2.2 Complexes with Organic Cations

An example of complexation that involves organic cations has been studied by Gutsche and coworkers.[47,48] Although chloroform solutions containing calixarenes and amines fail to give any indication of complex formation, acetonitrile, solutions behave quite differently. For example, a mixture of

[46] R. M. Noyes, *J. Am. Chem. Soc.*, **1962**, *84*, 513.
[47] L. J. Bauer and C. D. Gutsche, *J. Am. Chem. Soc.*, **1985**, *107*, 6063.
[48] C. D. Gutsche, M. Iqbal, and I. Alam, *J. Am. Chem. Soc.*, **1987**, *109*, 4314.

p-allylcalix[4]arene and *tert*-butylamine in acetonitrile shows shifts in the resonances of the *tert*-butyl hydrogens of the amine and the aryl hydrogens of the calixarene. Also, there is a sharpening of the methylene resonances as the ratio of amine to calixarene is increased, as illustrated in Figure 6.7. The effect on the amine can be duplicated by a mixture of *tert*-butylamine and picric acid, while that on the calixarene can be duplicated by a mixture of *p*-allylcalix[4]arene and sodium hydroxide. That the phenolic groups of the

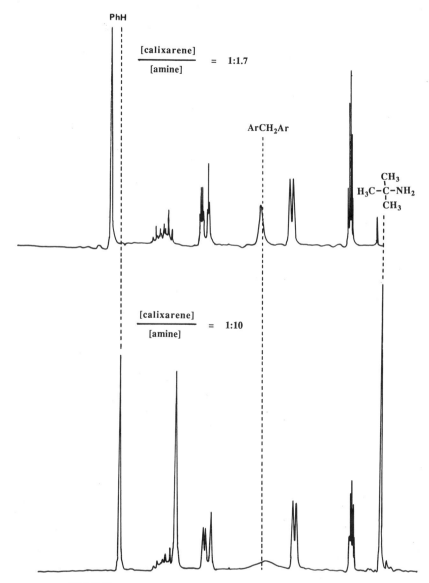

Figure 6.7   *¹H NMR spectra of mixtures of* tert-*butylamine and* p-*allylcalix[4]arene in acetonitrile-d₃ solution at* 30 °C[47]

calixarene play an integral part is also indicated by the much smaller perturbations that are observed in the ^{1}H NMR spectrum of a mixture of the tosylate of *p*-allylcalix[4]arene and *tert*-butylamine. These observations are explained in terms of a proton transfer from the calixarene to the amine to form a calixarene monoanion and an amine cation, followed by ion-pairing of the charged species to give an ion-pair complex, as shown in Figure 6.8. One piece of evidence in support of an *endo*-calix structure for the complex is the fact that the relaxation time (T_1) of the CH_3 protons of *tert*-butylamine is shorter in the amine–calixarene complex (0.79 s) than in a mixture of *tert*-butylamine and picric acid (1.21 s). Another piece of evidence is the variation in the conformational inversion barrier as a function of the particular amine that is used, *tert*-butylamine showing a larger conformational fixing effect than neopentyl amine as judged by the relative coalescence temperatures for the methylene resonances (see Chapter 4). This difference can be explained in terms of an *endo*-calix complex in which the preferred orientation of the amine places the amine hydrogens proximate to the oxygens in the bottom of the cavity, producing a 'tripod-like' association. In the *tert*-butylamine complex such an orientation places the methyl groups of the guest molecule comfortably in the middle of the calix, while, in the neopentylamine complex,

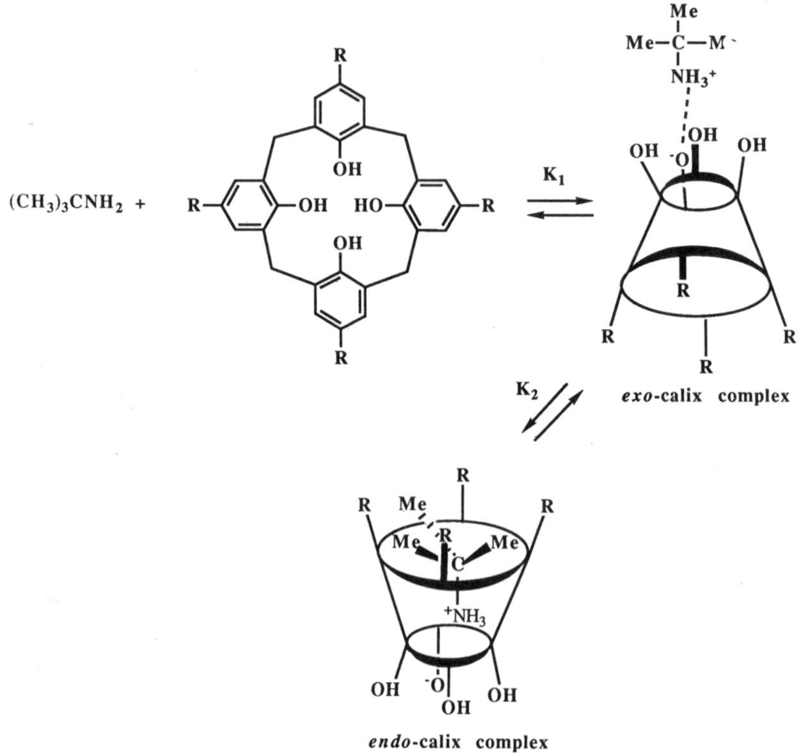

Figure 6.8   *Formation of* exo-calix *and* endo-calix *complexes from amines and calix[4]arenes*[47, 48]

the 'bend' in the neopentyl group tilts its *tert*-butyl portion against the side of the calix, making a 'tripod-like' association less effective. Probably the most compelling piece of evidence in support of an *endo*-calix complex comes from a two-dimensional Nuclear Overhauser Effect (2D-NOE) study on an equimolar mixture of *p*-allylcalix[4]arene and *tert*-butylamine in CD_3CN. Irradiation of the methyl protons of the *tert*-butylamine produces off-diagonal signals associated with the vinyl protons of the *p*-allyl substituents of the calixarene. Since the NOE effect is a through-space phenomenon that arises only if groups are proximate, it seems probable that the *tert*-butylamine is associated with the calixarene in a fashion that brings the methyl and allyl groups close to one another. The pathway for complex formation as depicted in Figure 6.8 has been studied in some detail by Gutsche and coworkers,[48] and values have been assigned to K_1 and K_2. The proton transfer step, K_1, can be measured essentially independently of the ion-pairing step, K_2, by noting the changes in the UV spectra of dilute solutions of host and guest (*ca.* 10^{-4} M). The ion-pairing step can be measured independently of the proton-transfer step by noting changes in the ^1H NMR spectra in more concentrated solutions (*ca.* 10^{-2} M) where the proton transfer is substantially complete. By the application of the Benesi–Hildebrand expression[49] to the concentration-dependent data collected in this fashion, the equilibrium constants are evaluated as $K_1 = ca.$ 5×10^4 and $K_2 = 50$—65, giving overall complexation constants in the range of 10^5—10^6.

6.2.3 Complexes of Neutral Organic Molecules

The interaction of calixarenes with toluene and chloroform has been studied by Bauer and Gutsche[47] who used the technique of aromatic solvent-induced shift (ASIS). With chloroform as the reference solvent, toluene as the putative guest molecule, and a linear tetramer as the reference compound, changes in chemical shifts of various protons in the solute calixarene were measured as a function of the chloroform:toluene ratio. An approximately linear relationship was observed, commensurate with (though not proving) a 1:1 complex, and from the slope of the line a K_{assoc} value of 1.1 was calculated. Although small, the association appears to be real, for other calixarenes behave in a different fashion. The ASIS value for *p-tert*-butylcalix[4]arene, for example, is considerably larger (*upfield* shift) than those for *p-tert*-butylcalix[6]arene and *p-tert*-butylcalix[8]arene, suggesting that the cyclic tetramer interacts with the toluene in a different manner from the larger cyclooligomers. Its ASIS value stands in particularly sharp contrast to that of the cyclic octamer, which shows a small *downfield* shift, indicating that toluene is less effectively oriented around the *tert*-butyl groups than in the reference compound (*i.e.* the linear oligomer of *p-tert*-butylphenol). This is commensurate with the solid state complexes of these oligomers, the cyclic tetramer forming a very tight complex with toluene and the cyclic octamer a very loose one (see p. 149).

---

[49] H. A. Benesi and J. H. Hildebrand, *J. Am. Chem. Soc.*, **1949**, *71*, 2703.

The features that the ASIS data reveal about the interaction between toluene and calixarenes can also explain some subtle differences in conformational behavior. For example, the conformational inversion barrier for *p*-(1,1,3,3-tetramethylbutyl)calix[4]arene is *ca.* 0.5 kcal/mole lower than that of *p-tert*-butyl or *p-tert*-pentylcalix[4]arene. That this is not simply a ponderal effect is indicated by the fact that the conformational inversion barrier for calix[4]arene itself is also lower by about the same amount. The ASIS experiments suggest that those calixarenes with the higher inversion barriers form *endo*-calix complexes with solvent molecules and that those with the lower inversion barriers do not, reinforcing the conclusions reached by the Parma group (see Section 6.1) that (a) unsubstituted calix[4]arene fails to form a complex, possibly because its cavity is too shallow, (b) *p*-(1,1,3,3-tetramethylbutyl)calix[4]arene fails to form an *endo*-calix complex because the *para*-substituents fold back into the calix, and (c) the *p-tert*-butyl and *p-tert*-pentylcalix[4]arenes form *endo*-calix complexes because the *para*-substituents extend the depth of the cavity but do not fold back into it.

The ASIS studies indicate that *p-tert*-butylcalix[4]arene forms complexes, albeit weak ones, with toluene and chloroform, but with little or no selectivity between the two. This is commensurate with the fact that both of these guests are tightly held by the solid calixarene. Attempts to demonstrate complex formation with other guests, however, have been unsuccessful, possibly because the putative guest compounds are unable to displace the chloroform or toluene from the calix to a sufficient extent to permit spectral detection. Compounds tested included anisole, nitrobenzene, *p*-xylene, bromobenzene, trichloromethylbenzene, trifluoromethylbenzene, acetone, *tert*-butylcyclohexanol, *p-tert*-butylphenol, benzonitrile, acetonitrile, trimethylacetonitrile, and phenylacetylene. The host molecules used were *p*-allylcalix[4]arene and its tetra-*p*-toluenesulfonate, the latter chosen because it is known to be frozen in a cone conformation and, therefore, to possess an enforced cavity. In the hope that complex formation might be demonstrated in a semi aqueous solvent, *p*-(2-hydroxyethyl)calix[4]arene and its tetra-*p*-toluenesulfonate ester were tested in 3:1 mixtures of acetonitrile:water or DMSO:water. Again, no measurable complex formation was observed, probably because the solvent molecules are too effectively complexed with the host.

Somewhat stronger solution-phase complexation with neutral molecules has been observed by Cram and coworkers[29] with the cavitand **5** (also *cf.* **14** in Figure 4.25 and **47** in Figure 5.14). The Benesi–Hildebrand method of analysis was applied to data obtained with 0.001 M solutions of **5** in $CDCl_3$ containing CS_2 (0.1—2 M), the changes in the ^1H NMR resonances of the calixarene protons being noted. The outward-facing (equatorial) methylene protons remain unaffected (± 0.04 ppm), while those on the inside of the cavity (axial) change appreciably; the aromatic protons on the 'lower rim' move upfield as much as 0.18 ppm from $\delta 7.35$, and the methylsilyl protons on the 'upper rim' move downfield as much as 0.40 ppm from $\delta - 0.55$. Complexation constants of 0.82, 8.1, and 13.2 for compounds **5a**, **5b**, and **5c**, respectively, were calculated. By measuring the K_{assoc} values as a function of

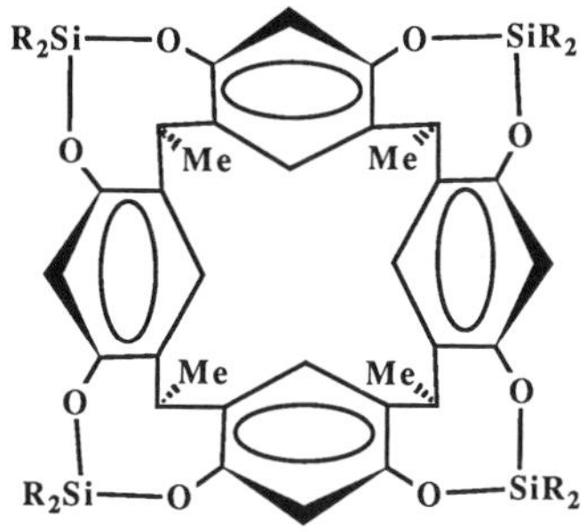

5a R = Me

5b R = Et

5c R = (CH$_2$)$_5$

temperature the thermodynamic parameters for **5a** at 212 °K were evaluated as $\Delta G° = -0.4$ kcal/mole, $\Delta H° = 3.5$ kcal/mole, and $T\Delta S = -3.1$ kcal/mole. Complexation was also demonstrated with CH$_3$C≡CH as the guest molecule but was not found with a number of other molecules, including benzene, iodine, methyl iodide, potassium ferricyanide, water, carbon dioxide, or methylene chloride in CDCl$_3$ solution. When a solution of **5a** in CDCl$_3$ is saturated with O$_2$ at 250 °K the resonances of the inward-facing protons broaden. The broadening was shown to be reversible by passage of N$_2$ through the solution, substantiating the postulate that a **5a–O$_2$** complex forms.

The resorcinol-derived calixarenes have been demonstrated by Aoyama and his coworkers[50] to be complexing agents for polar, water-soluble compounds. Compound **6**, prepared from resorcinol and dodecanal, was tested for its ability to extract glycerol, D-glucose, D-ribose, riboflavin, vitamin B$_{12}$, and hemin from aqueous solution into a CCl$_4$ or benzene solution containing the calixarene. While glycerol and D-glucose fail to be extracted, the other compounds all enter the CCl$_4$ phase to a significant extent. For example, the partition constants for riboflavin and vitamin B$_{12}$ between the organic phase (CCl$_4$) and the aqueous phase (1N-HCl) are 210 and 230, respectively, and hemin, which is a solid, can be solubilized in a benzene solution containing **6**.

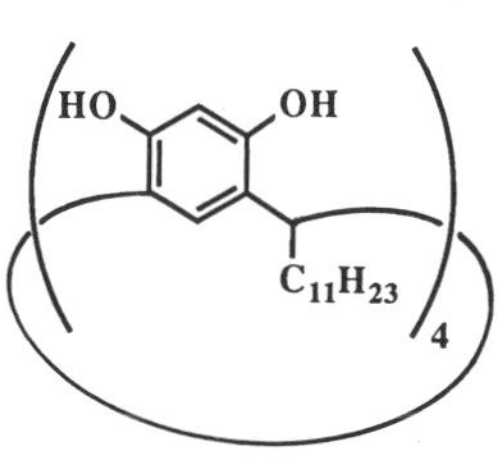

6

[50] Y. Aoyama, Y. Tanaka, H. Toi, and H. Ogoshi, *J. Am. Chem. Soc.*, **1988**, *110*, 634.

That hydrogen bonding must play a key role in this process is indicated by the failure of the octaacetate of **6** to extract any of these compounds. Hydrogen bonding is also indicated by the loss of fluorescence properties of riboflavin tetraacetate upon interaction with **6**, implicating the CONHCO moiety of the riboflavin in the formation of the complex. Although glycerol is not extracted from an aqueous solution, in neat form it interacts with **6** in the same fashion as does water to form a 4:1 complex. This leads the authors to suggest that the proximate OH groups at the 'upper rim' of the calixarene are the binding sites, operating individually in the case of small molecule guests such as glycerol or water or cooperatively in the case of larger molecule guests such as ribose whose complex is assigned the structure **6**-ribose:2 H_2O. Support for this concept is found in the fact that 4-dodecylresorcinol, which lacks proximate OH groups, is devoid of extraction activity.

The significant difference in the extractability of glucose and ribose persuaded Aoyama to pursue this facet in greater detail.[51] A number of sugars were tested, and it was found that they can be categorized as high-affinity sugars (*e.g.* ribose and 2-deoxyribose), moderate-affinity sugars (*e.g.* erythrose and arabinose), low-affinity sugars (*e.g.* glucose and mannose), and no-affinity sugars (*e.g.* xylose and lyxose). Correlating the extractability of these sugars with the extent to which they exist in the furanose form, he concludes that the primary requisite for strong complexation is a furanose ring attached to a CH_2OH group: thus, 2-deoxyribose (which exists 20% in the furanose–CH_2OH form) is extracted more effectively than glucose (which exists 0.2% in the furanose–CH_2OH form); glucose, in turn, is more effectively extracted than sorbitol (which cannot exist in a cyclic form).

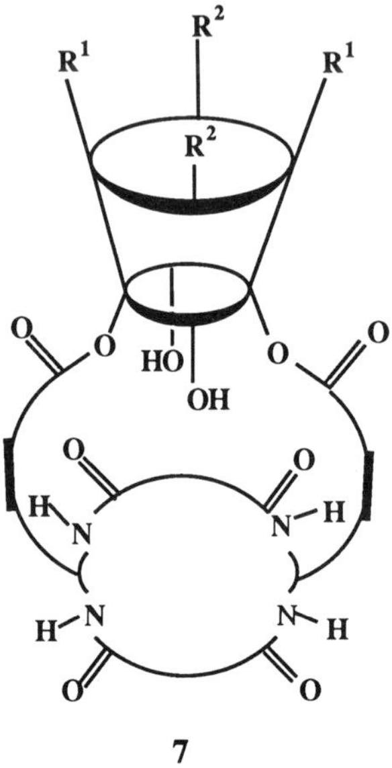

7

---

[51] Y. Aoyama, S. Sugahara, and Y. Tanaka, submitted for publication.

Table 6.4 Complexation constants with a 'double cavity' calixarene (7)[52]

Phenols	K_{assoc}/M	Carboxylic acids	K_{assoc}/M	Aromatic amines	K_{assoc}/M	Aliphatic amines	K_{assoc}/M	Aromatic–aliphatic amines	K_{assoc}/M
4-NO$_2$	55	ICH$_2$CO$_2$H	9	Pyridine	33	(CH$_3$)$_2$CHCH$_2$CH$_2$NH$_2$	13	p-MeO-benzyl amine	19
3-NO$_2$	40	Cl$_2$CHCO$_2$H	6	4-CH$_3$-pyridine	6	CH$_3$CH$_2$CH$_2$CH$_2$NH$_2$	12	p-MeO-phenylethyl amine	16
4-CN	31	ClCH$_2$CH$_2$CO$_2$H	5	4-CN-pyridine	0	(CH$_3$)$_2$CHNH$_2$	13	p-MeO-phenylpropyl amine	10
4-CF$_3$	21	BrCH$_2$CO$_2$H	5	2,6-di-CH$_3$-pyridine	0	CH$_3$CH$_2$CH(CH$_3$)NH$_2$	0		
4-Br	21	Br$_2$CHCO$_2$H	5	Imidazole	16	(CH$_3$)$_3$CCH$_2$NH$_2$	0		
2-NO$_2$	0					(CH$_3$CH$_2$)$_2$NH	0		
2-Br	0					CH$_3$CH(CH$_2$OH)NH$_2$	20		
2,4-di-NO$_2$	0					HOCH$_2$CH$_2$NH$_2$	15		
4-NO$_2$-3,5-di-CH$_3$	0					CH$_3$CH(OH)CH$_2$NH$_2$	11		

Among the strongest of the complexes of calixarenes with neutral molecules are those that have been prepared by Gutsche and See[52] using the 'double cavity' calixarene (**7**) described in Chapter 5 (see **17** in Figure 5.6). This compound appears to behave both as a hydrogen bond donor, forming complexes with amines, and as a hydrogen bond acceptor, forming complexes with acids and phenols. By measuring the ¹H NMR shifts that occur upon mixing the 'double cavity calixarene' with the various guest molecules and by applying a modified Benesi–Hildebrand expression to the data, the K_{assoc} constants shown in Table 6.4 are obtained. That acidity and basicity, *per se*, play some part in determining the strength of the complex is indicated by the K_{assoc} values of 4-nitrophenol *vs.* 4-bromophenol and pyridine *vs.* 4-cyanopyridine. Shape and bulk, however, play an even greater role. Thus, 3-nitrophenol and 4-nitrophenol form strong complexes, but 2-nitrophenol, 2-bromophenol, and 2,4-dinitrophenol all fail to do so. Similarly, *n*-butylamine, isopropylamine, and isobutylamine form moderately strong complexes, but *sec*-butylamine, *tert*-butylamine, and neopentylamine all fail to do so. Since neither the free host **7** nor any of its complexes have yet yielded to *X*-ray analysis, the geometry of complexation in this series remains to be established. The 'double cavity' calixarene (see CPK models in Figure 6.9) is unusual in that it possesses both an enforced cavity on the 'lower rim' and a semi-flexible cavity on the 'upper rim'. The two phenolic moieties can swing on their *meta–meta* axes, creating a full calix in one position or closing the side portals of the lower cavity in another position.

Figure 6.9   *Space-filling molecular models of a double cavity calixarene*

⁵² C. D. Gutsche and K. A. See, to be published.

6.3 Solution State Complexes of Water-soluble Calixarenes

6.3.1 Metal Cations as Guests

References to the work of Shinkai appear in numerous places throughout this book. Although a relative latecomer to the field, he has already made important contributions that have taken calixarene chemistry in new directions. Seiji Shinkai was born in Fukuoka, Japan in 1944 and received his PhD in 1972 from Kyushu University under the direction of Toyoki Kunitake. The next two years were spent in the laboratories of Thomas Bruice at the University of California, Santa Barbara, following which he returned to his native country to take a faculty position at his *alma mater*. After only a year he moved to the University of Nagasaki where he remained until 1987, returning that year again to the University of Kyushu. Bringing to calixarene chemistry a viewpoint that reflects the influence of this training in physical organic chemistry, he has embarked on an ambitious program focusing on the uses to which calixarenes can be put. A good example, and one that represents an especially interesting study of cation complexation by calixarenes, deals with the extraction of uranium from sea water. Uranium, present in sea water as the UO_2^{2+} cation (strongly complexed with CO_2), forms complexes possessing a pseudoplanar pentacoordinate or hexacoordinate geometry. Taking cognizance of this, Shinkai *et al.*[53,54] tested several water-

Seiji Shinkai

[53] S. Shinkai, H. Koreishi, K. Ueda, and O. Manabe, *J. Chem. Soc., Chem. Commun.*, **1986**, 233.
[54] S. Shinkai, H. Koreishi, K. Ueda, T. Arimura, and O. Manabe, *J. Am. Chem. Soc.*, **1987**, *109*, 6371.

soluble sulfonatocalixarenes (*cf.* compound **22** in Figure 5.8). Although the
p-sulfonatocalix[4]arenes have little complexing ability for UO_2^{2+}, the data in
Table 6.5 show that both the *p*-sulfonatocalix[5]arenes and *p*-sulf-
onatocalix[6]arenes have high K_{assoc} values and dramatic selectivities for
UO_2^{2+} in preference to other cations such as Ni^{2+}, Zn^{2+}, and Cu^{2+}. The pH-
dependence of the absorption spectra of these solutions indicates that the calixarenes form stable 1:1 complexes in basic solution and even in neutral
solution. The dramatic difference in log *K* values between the calix[4]arene
on the one hand and the calix[5]arenes and calix[6]arenes on the other is attributed either to (a) the UO_2^{2+} cation exactly fitting the cavity of the larger
calixarenes but not the smaller, *i.e.* a 'hole-size selectivity' or (b) the larger
calixarenes being better able to provide the ligand groups arranged in the
fashion required for pseudoplanar penta- or hexa-coordination on the edge
of the calixarenes, *i.e.* a 'coordination-geometry selectivity'. Shinkai *et al.*
consider the second explanation to be the more likely, because the cavity size
of calixarenes is quite variable as a result of their conformational flexibility.

Table 6.5   *log* K_{assoc} *values for complex formation of sulfonatocalixarenes
with* UO_2^{2+}, Mg^{2+}, Ni^{2+}, Zn^{2+}, *and* Cu^{2+} [53,54]

R	n	UO_2^{2+}	Mg^{2+}	Ni^{2+}	Zn^{2+}	Cu^{2+}
H	4	3.2				
CH_2CO_2H	4	3.1				
H	5	18.9 ± 0.6				
CH_2CO_2H	5	18.4 ± 0.1				
H	6	19.2 ± 0.1	small	2.2	5.5	8.6
CH_2CO_2H	6	18.7 ± 0.1	small	3.2	5.6	6.7
Me	6	3.2 ± 0.2				

From the ratios of the K_{assoc} values for UO_2^{2+} relative to the other cations
in Table 6.5 the following selectivity factors can be calculated: $> 10^{17}$ for
Mg^{2+}, 10^{17} and $10^{15.2}$ for Ni^{2+}, $10^{13.7}$ and $10^{13.1}$ for Zn^{2+}, and $10^{10.6}$ and $10^{12.0}$
for Cu^{2+}. These large values are surprising in view of the fact that the dia-
meters of all five divalent species are very similar, ranging from 0.66—0.87 Å.
The striking advantage possessed by UO_2^{2+} is attributed to the geometric
requirements of its complexation, *viz.* pseudoplanar penta- or hexacoordinate, in contrast to that of the four other cations, *viz.* square-planar or
tetrahedral. It is argued that the semi-rigid character of the calixarenes is
conducive to the maintenance of the pseudoplanar geometry required for
UO_2^{2+} complexation.

6.3.2   Organic Cations as Guests

Taking their cue from the work of Gokel and Cram in which it was found that
crown ethers with the appropriate dimensions can solubilize arenediazonium

salts,[55] Shinkai's group[56,57] has studied the interaction of sulfonato-calix[6]arenes with these cations. By measuring the rates of the decomposition of the benzenediazonium salts in the presence of the calixarene (**8**), the complexation constants were measured to be *ca.* 10^2 M^{-1} for **8a**, 4.2×10^6 M^{-1} for **8c**, and 5.6×10^4 M^{-1} for **8d**. To gain insight into the mode of binding of the aryldiazonium salt in the calixarene cavity, the interaction with 4-(4′-dimethylaminophenylazo)benzenediazonium tetrafluoroborate (**9**) was explored. While **8a** fails to perturb the absorption of **9** at 612 nm in aqueous solution, **8b** shifts it to 609 nm and **8c** to 595 nm, indicating that the diazonium compounds are bound in the relatively hydrophobic site at the 'lower rim' of the calixarene. This system is discussed in more detail in the next chapter.

8a   R = H

8b   R = Me

8c   R = n-hexyl

8d   R = n-dodecyl

8e   R = CH$_2$CHCH$_2$CH$_3$
          |
          CH$_3$

Another organic cation used effectively by the Shinkai group in studying the complexing ability of *p*-sulfonatocalixarenes is trimethylanilinium.[58] By measuring the magnitude of the upfield shifts in the ^{1}H NMR resonances of this guest as a function of the ratio of host to guest, they calculated K_{assoc} values. By measuring the K_{assoc} values over a temperature range of 0–80 °C they obtained the thermodynamic parameters shown in Table 6.6. These reveal some interesting variations. Complex formation with the cyclic tetramer is driven by a favorable enthalpy change, whereas with the cyclic hexamer and octamer it is driven by an entropy increase. This difference is attributed to stronger electrostatic interactions in the calix[4]arene and stronger hydrophobic interactions in the larger calixarenes which are better able to accommodate the guest within their cavities. The 10-fold lower association constant for the cyclic hexamer, though, is surprising. These data are also interesting in showing a biphasic behavior for the cyclic octamer. This is interpreted as an initial formation of a 1:1 complex followed by acquisition of a second guest molecule to form a 2:1 complex. The almost identical values of K_1 and K_2 led the authors to invoke the 'pinched' conformation[47] in which the flexible ring puckers in a fashion that creates two

[55] G. W. Gokel and D. J. Cram, *J. Chem. Soc., Chem. Commun.*, **1973**, 481.

[56] S. Shinkai, S. Mori, K. Araki, and O. Manabe, *Bull. Chem. Soc. Jpn.*, **1987**, *60*, 3679.

[57] S. Shinkai, S. Mori, T. Arimura, and O. Manabe, *J. Chem. Soc., Chem. Commun.*, **1987**, 238.

[58] S. Shinkai, K. Araki, and O. Manabe, *J. Am. Chem. Soc.*, **1988**, *110*, 7214.

Table 6.6   *Association constants and thermodynamic parameters for the
complexes of* p-*sulfonatocalix[n]arenes and trimethylanilinium
cation*

p-Sulfonato- calix[n]arene	K_{assoc}/M	$\Delta G°$/kcal mole^{-1}	$\Delta H°$/kcal mole^{-1}	$\Delta S°$/e.u.
$n = 4$	5600 ± 250	-5.1 ± 0.5	-6.1 ± 0.3	-3.6 ± 0.8
$n = 6$	550 ± 40	-3.7 ± 0.2	-0.25 ± 0.10	11.7 ± 0.8
K_1	5200 ± 50	-5.1 ± 0.2	0.0 ± 0.10	17.0 ± 0.3
$n = 8$				
K_2	4600 ± 50	-5.0 ± 0.1	0.00 ± 0.10	16.7 ± 0.1

cyclic tetramer-like cavities. Although the thermodynamic parameters
provide good support for this idea, it is also possible that the two guest
molecules are occluded side by side in the cavity of the cyclic octamer in its
'pleated loop' conformation. Still another interesting observation coming out
of this study is the fact that the organic cation trimethylanilinium stabilizes
the cone conformation of the cyclic tetramer more effectively than do
inorganic cations.

A complexation phenomenon that bears some resemblance to that of
amines with *p*-alkylcalixarenes (see Figure 6.9) is the interaction of *p*-
sulfonatocalixarenes with Phenol Blue, both processes involving the transfer
of a proton as an integral step. Phenol Blue, introduced many years ago as an
indicator for solvent polarity,[59] has an absorption at *ca.* 660 nm in water
which shifts to shorter wavelengths in nonpolar solvents (*e.g.* 552 nm in
cyclohexane), the blue shift being attributed to destabilization of the charge-
separated excited state. Shinkai and coworkers[60] discovered the surprising
fact that Phenol Blue in the presence of *p*-sulfonatocalix[6]arene (**8a**) shifts its
absorption maximum to *longer* wavelength (685 nm), suggesting that the
calixarene provides an environment more polar than water. They rationalize
this by postulating that the calixarene stabilizes the charge-separated excited
state of Phenol Blue by forming the *endo*-calix complex **10**, as shown in
Figure 6.10, in which electrostatic interactions play a dominant role. The
ethers of *p*-sulfonatocalix[6]arene **8b** and **8d**, on the other hand, behave differently, **8b** showing a very small red shift to 662 nm and **8d** showing a strong
biphasic blue shift to 603 nm at high concentration and 592 nm at low concentration. The smaller blue shift with **8d** that occurs at high concentration is
similar to that induced by sodium dodecyl sulfate micelles and is attributed to
the incorporation of the Phenol Blue in a hydrophobic pocket in the calixarene. The larger blue shift with **8d** that occurs at low concentration is
ascribed[57] to 'protonation' of Phenol Blue, since the same shift can be

[59] L. G. S. Brooker and R. H. Sprague, *J. Am. Chem. Soc.*, **1941**, *63*, 3214.
[60] S. Shinkai, S. Mori, H. Koreishi, T. Tsubaki, and O. Manabe, *J. Am. Chem. Soc.*, **1986**, *108*,
2409.

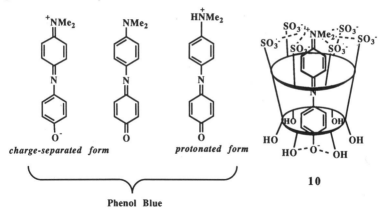

Figure 6.10 *Complexation of* p*-sulfonatocalix[6]arene with Phenol Blue*

induced with strong acid in the absence of the calixarene; furthermore, the effect disappears in a buffered solution at pH 8.7. Protonated Phenol Blue is postulated to form an *endo*-calix complex comparable to that shown in Figure 6.10 for **10** in which the positive charge is stabilized by negative charges on the proximate sulfonato groups at the 'upper rim' of the calix. The K_{assoc} values for the complexes with Phenol Blue are reported[60] to be $5.6 \times 10^2 \ M^{-1}$ for **8a** and $2.0 \times 10^5 \ M^{-1}$ for **8d**.

Although none of the phenol-derived calixarenes have yet been tested for complexation with organic ammonium ions, a resorcinol-derived calixarene has been shown by Hans-Jörg Schneider and coworkers[61] at the University of Saarbrucken to form complexes with several such species. The all-*cis* isomer of C-methylcalix[4]resorcinarene, which exists in the flattened cone conformation as a neutral species (see structure A_{aaaa} in Figure 4.9), assumes a full cone conformation when it is transformed to the water-soluble tetraanion. The tetraanion interacts with a number of ammonium compounds as well as with a neutral compound to give complexes with K_{assoc} values ranging from less than 10 to greater than 10^4, the $\Delta G°$ values for which are given in Figure 6.11. Inspection of these data shows that the magnitude of $\Delta G°$ is a function of the distance between the positive charge on the guest molecule and the anionic charges on the 'upper rim' of the host molecule, the greater the separation the weaker is the complex. If the charge becomes too far removed or if it is not present, as in the case of the neutral amino alcohol, the K_{assoc} values fall to less than $10 \ M^{-1}$. That the driving force for complexation derives primarily from electrostatic attraction was demonstrated by a study of the effect of solvent composition on the K_{assoc} values.[62] The *trans,cis,trans,cis*

[61] H.-J. Schneider, D. Güttes, and U. Schneider, *Angew. Chem., Int. Ed. Engl.*, **1986**, *25*, 647; idem., *J. Am. Chem. Soc.*, **1988**, *110*, 6449.
[62] H.-J. Schneider, R. Kramer, S. Simova, and U. Schneider, *J. Am. Chem. Soc.*, **1988**, *110*, 6442.

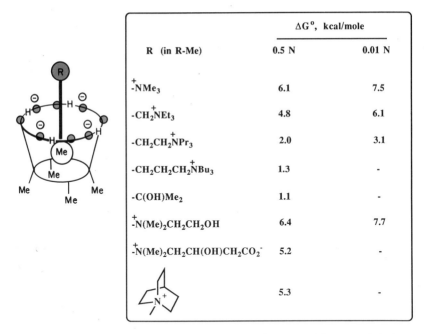

R (in R-Me)	ΔG°, kcal/mole	
	0.5 N	0.01 N
$-\overset{+}{N}Me_3$	6.1	7.5
$-CH_2\overset{+}{N}Et_3$	4.8	6.1
$-CH_2CH_2\overset{+}{N}Pr_3$	2.0	3.1
$-CH_2CH_2CH_2\overset{+}{N}Bu_3$	1.3	-
$-C(OH)Me_2$	1.1	-
$-\overset{+}{N}(Me)_2CH_2CH_2OH$	6.4	7.7
$-\overset{+}{N}(Me)_2CH_2CH(OH)CH_2CO_2^-$	5.2	-
	5.3	-

Figure 6.11 *Complexes of a resorcinol-derived calixarene with ammonium ions and an amine alcohol[61]*

isomer of C-methylcalix[4]resorcinarene, which exists in a flattened partial cone conformation as a neutral species (see structure D_{aaaa} in Figure 4.9), assumes a partial cone conformation when it is transformed to the dianion, and it shows complexing ability toward ammonium ions. In contrast to the all-*cis* isomer, which loses only four protons even in the presence of NaOMe, the *trans,cis,trans,cis* isomer loses all eight of its OH protons under these conditions and forms an octaanion. The octaanion shows no complexing ability toward ammonium ions, and it is postulated that, like the neutral species, it exists in a flattened partial cone conformation that is essentially devoid of a cavity.

6.3.3   Organic Anions as Guests

Circular dichroism, so successfully exploited in studying the complexation behavior of the cyclodextrins, has been introduced into calixarene chemistry by Shinkai and coworkers. In one study[63] they employed the technique of induced circular dichroism by using *p*-nitrocalix[6]arene as a sparingly water soluble, achiral host and the binaphthyl phosphate **11** as the chiral guest. Significant elipticities were observed with *p*-nitrocalix[6]arene and *p*-nitrocalix[8]arene but none with *p*-nitrocalix[4]arene. This was interpreted as

[63] T. Arimura, S. Edamitsu, S. Shinkai, O. Manabe, T. Muramatsu, and M. Tashiro, *Chem. Lett.*, **1987**, 2269.

11

indicating the ability of the guest to engage the two larger calixarenes in an *endo*-calix complex while rejecting the smallest calixarene. That these are 1:1 complexes was demonstrated by a 'continuous variation' method in which the elipticity is plotted as a function of the ratio [calixarene]:{[calixarene]+[guest]}, a maximum at 0.5 being taken to indicate a 1:1 complex.

6.3.4 Neutral Molecules as Guests

The complexation properties of the *p*-sulfonatocalix[6]arenes with a variety of guest molecules in addition to Phenol Blue have been carefully investigated by Shinkai. To evaluate the aggregation characteristics of the ethers of these calixarenes a variety of techniques were employed,[62] including light-scattering, surface tension, conductance, and fluorescence measurements. It was determined that **8c** has a critical micelle concentration (cmc) of *ca.* 6×10^{-4} M, while **8d** fails to show a detectable cmc. This behavior of **8d** is interpreted in terms of a 'unimolecular micelle', based on the different ways in which the two compounds interact with small molecules such as pyrene, 2-anilinonaphthalene, and the dye Orange OT. Plots of the amount of Orange OT carried into the aqueous phase by the host *vs.* the concentration of the host reveal that **8c** acts very much like sodium dodecyl sulfate, a typical micelle-forming compound. It requires *ca.* 19 molecules of **8c** per molecule

Orange OT

of Orange OT to give a complex with $K_{assoc} = 2.3 \times 10^7$. Compound **8d**, on the other hand, behaves differently and requires only 1.6 molecules of **8d** per molecule of Orange OT to give a complex with $K_{assoc} = 5.2 \times 10^5$ M^{-1}. The K_{assoc} values for **8c** and **8d** with 2-anilinonaphthalene was measured by fluorescence spectrometry and found to be 8.3×10^5 and 3.4×10^5 M^{-1}, respectively. The characterization of **8d** as a 'unimolecular micelle' is further substantiated by the fact that the solubilizing ability of **8d** toward Orange OT increases with increasing temperature, whereas that of SDS decreases.

Fluorescence measurements with 2-anilinonaphthalene (AN) as the guest also support the idea that **8d** forms a 1:1 complex. Thus, **8a** and **8b** have no effect on the fluorescence of AN, but **8d** shifts it from 447 to 420 nm. Its spectral behavior in aqueous solution shows saturation at a concentration (3×10^{-5}) identical with that of AN.

Another facet of Shinkai's use of circular dichroism as an analytical device for the study of complexation involves the host molecule **8e** in which chiral alkyl groups are affixed to the phenolic oxygens.[59] Using **8e** at a concentration below the cmc $(1.1 \times 10^{-3}$ M), the induced circular dichroism observed in the guest molecules **12a** and **12b** was interpreted as indicating inclusion of the guests in a cavity of the calixarene close to the chiral centers of the *S*-2-methylbutyl groups.

$$\text{Et}_2\text{N}-\!\!\!\left\langle\bigcirc\right\rangle\!\!\!-\text{N}=\text{N}-\!\!\!\left\langle\bigcirc\right\rangle\!\!\!-\text{R}$$

12a   R = NO$_2$

12b   R = CN

In contrast to *S*-*p*-(methylbutoxy)-benzenesulfonic acid which shows no CD spectrum, **8e** shows a strong CD spectrum in H$_2$O–DMF solution. This is interpreted[64] in terms of an 'alternate' conformation which brings the aromatic rings in **8e** proximate to the chiral centers, a conclusion that finds further corroboration in the sharp singlet arising from the methylene resonances in the ^{1}H NMR spectrum. Changes in the CD spectrum in the presence of alkanols makes possible the measurement of the following K_{assoc} values: 1-hexanol (1.4×10^2), 1-heptanol (1.2×10^3), 1-octanol (7.8×10^3), and cyclohexanol (0.8×10^2). Concomitant with the changes in the CD spectrum of the **8e**–octanol complex are changes in the ^{1}H NMR spectrum; the sharp singlet from the methylene protons characteristic of the 'alternate' conformation gives way to a much broader resonance, and the singlet from the aromatic protons changes at 0 °C into a pair of lines. The authors postulate that the effect of the inclusion of the guest molecule is to change the shape of the calix[6]arene from an 'alternate' to a 'cone' conformation. An aqueous phase complexation study embracing calixarenes of five different sizes, ranging from the calix[4]arene through the calix[8]arene, has been carried out by Gutsche and Alam[65] with *p*-(diallylaminomethyl)calixarenes (**13**) and *p*-(carboxyethyl)calixarenes(**14**), prepared as described in Chapter 5 (see Figure 5.12 and 5.13). By means of a solid–liquid extraction technique that has been well described by Diederich[66] the K_{assoc} values were obtained for a series of aromatic hydrocarbons of various dimensions, as shown in Table 6.7. The data in Table 6.7 show that in those cases where K_{assoc} is above

[64] S. Shinkai, T. Arimura, H. Satoh, and O. Manabe, *J. Chem. Soc., Chem. Commun.*, **1987**, 1495.

[65] C. D. Gutsche and I. Alam, *Tetrahedron*, **1988**, 4689.

[66] F. Diederich and K. Dick, *J. Am. Chem. Soc.*, **1984**, *106*, 8024.

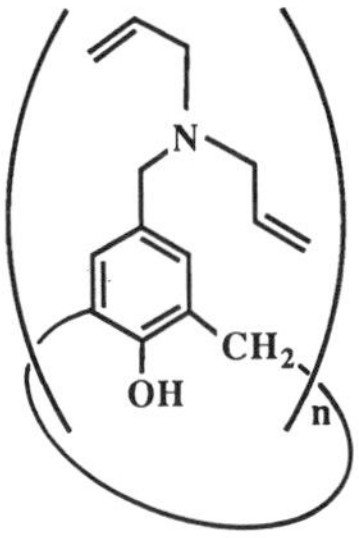

13 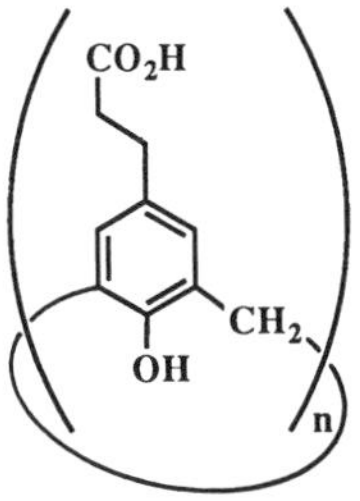   14

the threshold value of *ca.* 10^2 M^{-1}, it varies over a 25-fold range from a low of 6×10^2 M^{-1} to a high of 1.5×10^4 M^{-1}. It is noteworthy that for a given guest molecule the values for the aminomethylcalixarenes and the carboxyethylcalixarenes are approximately the same, suggesting that the site of complexation is not close to the water solubilizing groups but is at the hydroxyl array on the 'lower rim'. The working premise of the authors is that complexation occurs by insertion of a portion of the aromatic hydrocarbon into the 'lower rim' of the calixarene. Measurements on CPK models give the dimensions for the aromatic hydrocarbons shown in Figure 6.12 and the following approximate dimensions for the eliptical 'lower rim' annuli of the calixarenes:

calix[4]arene — *ca.* 2 Å diameter
calix[5]arene — 3 Å × 6.2 Å
calix[6]arene — 3 Å × 7.6 Å
calix[7]arene — 3 Å × 8.6 Å
calix[8]arene — 3 Å × 11.7 Å

According to these dimensions the calix[4]arenes should be too small to accept naphthalene or durene, and this accords with the observations. Calix[8]arenes, on the other hand, should be able to accept the larger molecules such as perylene and this, too, accords with the observations and indicates that there is a rough correlation between dimensional complementarity and complex formation in these systems. There are some unexplained anomalies, however, such as the failure of durene to form a complex with calix[6]arenes, although the size of the host moelcules appears to be adequate, and the failure of coronene and decacyclene to form complexes with calix[8]arene.

6.4   Concluding Remarks

Complexation phenomena, which have played an important role throughout the history of chemistry, came into particular prominence for organic chemists about 30 years ago with the advent of the cyclodextrins and then the crown ethers. Complexation studies with the cyclodextrins have been directed primarily to nonionic molecular guests, while those with the crown ethers have been mostly concerned with ionic guests. Calixarenes add to the growing list of complex-forming macrocyclics, and it is interesting to note

Table 6.7 *Association constants, K_{assoc}/M for p-(diallylaminomethyl)calixarenes (13) in 0.01 M HCl and p-(carboxyethyl)calixarenes (14) in 0.01 M K_2CO_3 with aromatic hydrocarbons in aqueous solution: A – durene, B – naphthalene, C – anthracene, D – phenanthrene, E – fluoranthene, F – pyrene, G – perylene, H – coronene, I – decacyclene[65]*

Guest

Host	A	B	C	D	E	F	G	H&I
13 ($n=4$)	0	0	0	0	0	0	0	0
14 ($n=4$)	0	0	0	0	0	0	0	0
13 ($n=5$)	0	3.3×10^3	9.0×10^3	0	0	$< 10^2$	$< 10^2$	0
14 ($n=5$)	0	3.7×10^3	9.1×10^3	0	4.0×10^3	$< 10^2$	$< 10^2$	0
13 ($n=6$)	0	4.5×10^3	1.6×10^3	1.9×10^3	4.0×10^3	$< 10^2$	$< 10^2$	0
14 ($n=6$)	0	3.7×10^3	3.0×10^4	3.0×10^3	3.4×10^3	$< 10^2$	$< 10^2$	0
13 ($n=7$)	3.9×10^3	3.0×10^3	8.3×10^3	8.9×10^3	3.0×10^3	9.0×10^3	1.0×10^4	0
14 ($n=7$)	3.9×10^3	3.9×10^3	1.1×10^4	8.9×10^3	3.6×10^3	1.1×10^4	9.0×10^3	0
13 ($n=8$)	2.8×10^3	1.1×10^3	5.0×10^3	4.1×10^3	1.5×10^4	6.0×10^4	1.0×10^4	0
14 ($n=8$)	2.8×10^3	6.1×10^2	9.4×10^3	4.1×10^3	1.4×10^4	4.0×10^4	4.0×10^3	0

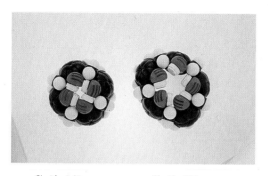

Calix[4]arene   Calix[5]arene

Calix[6]arene   Calix[7]arene

Calix[8]arene

CPK Models of calixarenes from the hydroxyl face

Naphthalene   Phenanthrene   Pyrene

Fluoranthene   Perylene

CPK Models of aromatic hydrocarbons

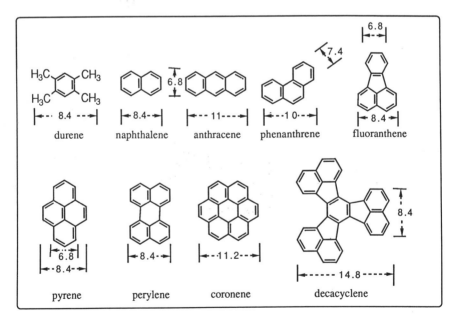

Figure 6.12 *Structure and dimensions (in Å) of the aromatic hydrocarbons used in the complexation with **13** and **14** measured from CPK models. The thickness in all cases is* ca. *3 Å except for durene in which it is* ca. *3.6 Å*

that they are being explored as hosts for both of these varieties of guests. They possess cation complexing capabilities that in some cases rival those of crown ethers, and they show neutral molecule complexing capabilities that in some respects resemble those of the cyclodextrins. As more highly functionalized and contoured calixarenes become available through the efforts of the synthetic chemists, new and interesting complexation characteristics will undoubtedly unfold.

CHAPTER 7

Using the Baskets: Calixarenes in Action

'The used key is always bright'

Benjamin Franklin, *The Way to Wealth*

Chemical research is sometimes an art-form more than a formal science, particularly as practiced in academic environments. In contrast to the mission orientation that motivates most industrial research, the chemist in the university laboratory enjoys the freedom to wander and wonder.[1] Calixarene chemistry provides a curious and interesting blend of these two worlds of scientific research. Although the birth of calixarene chemistry occurred in an academic setting, it was the outcome of an investigation whose goal was to gain insight into the manufacture of Bakelite, one of the world's most profitable chemical commodities. Incorporated some years later, albeit unwittingly, into a different kind of industrial process by the Petrolite Corporation, it subsequently traversed the academic–industrial interface in St Louis to re-emerge as an integral part of a bioorganic chemistry research program. Elsewhere in the world calixarenes were simultaneously re-emerging as part of a polymer-oriented research program in Mainz and a crown ether-oriented research program in Parma. During the decade that followed, calixarene chemistry progressed steadily, if haltingly, in these academic laboratories, ultimately becoming visible to a wider community of researchers that included not only academics but also individuals and institutions with entrepreneurial aspirations. Chapters 2—6 of this book deal primarily with the art-form aspects of calixarene chemistry in focusing on the methods that have been devised for synthesizing, contouring, and functionalizing these molecular baskets. In this last chapter attention is now turned to some of the ways in which the baskets have been, or might be, put to use. The ability of the calixarenes to get into action and do something, whether it be truly useful in the utilitarian sense or just intriguing in an intellectual sense, provides the central focus.

---

[1] 'A good way to tell how the work is going is to listen in the corridors. If you hear the word, "Impossible!" spoken as an expletive, followed by laughter, you will known that someone's orderly research plan is coming along nicely', Lewis Thomas in 'Lives of a Cell'.

7.1 The Patent Literature of Calixarenes

Patents are generally sought for items thought to have utilitarian promise, and the extent of the patent literature for a particular item gives some measure of its perceived utility. Almost two dozen patents have been issued during the last several years describing a variety of calixarenes, claiming various uses, and suggesting that these compounds may have a place in the world of commerce. A number of these patents focus on the calixarenes as ionophores, a feature that was discovered by Izatt and coworkers to whom a patent has been issued for a process that uses this technology for recovering cesium from radioactive wastes.[2] The Loctite Corporation has been particularly active in securing patents that feature calixarenes as an ingredient in their cyanoacrylate adhesives.[3-7] One of these patents[3] claims calixarenes to be useful not only as ion exchangers but also as phase-transfer catalysts, coatings, adhesives and potting compounds. Another[6] claims calixarenes to be useful as accelerators for cyanoacrylate adhesives for wood, leather, ceramic, plastics, and metals, and it states that 'adhesives containing a calixarene of the general structure **1** substantially reduce the cure times of

1

deactivating substrates such as wood'. The attachment of calixarenes to polymer supports has been explored in considerable detail by the Loctite chemists[8] who have used both 'lower rim' and 'upper rim' modes of attachment. Two examples of the former are illustrated in Figure 7.1, including (a) use of phenolic hydroxyl groups as points of attachment to a mixture of toluene diisocyanate and polyethylene glycol and (b) use of allyloxy moieties for platinum-catalyzed addition to dithiols. Two examples of the latter are

[2] R. M. Izatt, J. J. Christensen, and R. T. Hawkins, U.S. Patent 4,477,377, 16 Oct **1984** (*Chem. Abstr.*, 101: 21833m).

[3] S. J. Harris, M. A. McKervey, D. P. Melody, J. Woods, and J. M. Rooney, Eur. Pat. Appl. EP 151,527, **1984** (*Chem. Abstr.*, 103: 216392x).

[4] S. J. Harris, J. M. Rooney, and J. G. Woods, Eur. Pat. App. EP 196, 895, **1986** (*Chem. Abstr.*, 106: 67887v).

[5] M. A. McKervey, U.S. Patent 4,622,414, **1986.**

[6] R. G. Leonard and S. J. Harris, U.S. Patent, 4,695,615, **1987.**

[7] B. Kneafsey, J. M. Rooney, and S. J. Harris, Brit. U.K. G.B. 2,185,261, **1987** (*Chem. Abstr.*, 108: 95105y).

[8] S. J. Harris, J. G. Woods, and J. M. Rooney (Loctite, Ireland, Ltd.), U.S. Patent 4,642,362, **1987.**

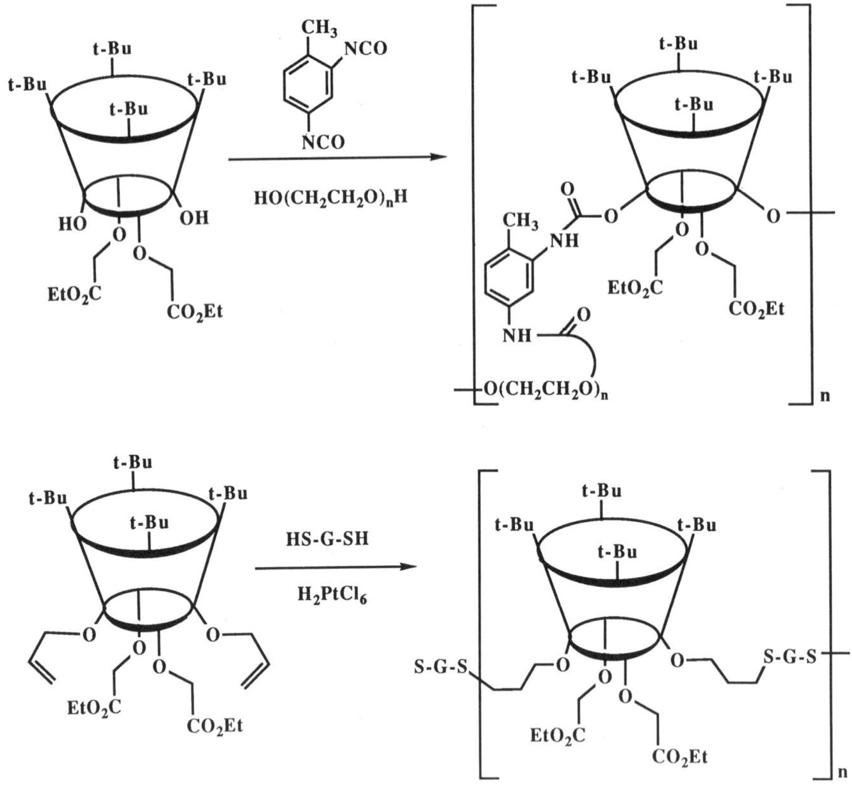

Figure 7.1 *Polymeric calixarenes* via *'lower rim' attachement[8]*

illustrated in Figure 7.2, including (a) use of the free *p*-positions for AlCl₃-catalyzed reaction with chloromethylated polystyrene and (b) use of *p*-allyl moieties for platinum-catalyzed addition to dithiols.

Eight patents were issued to the Hitachi Chemical Co.[9] in 1984 involving the preparation of calixarenes from *p*-phenylphenol, *p*-(4-R-phenyl)phenyl-phenol (R not specified), and *p*-R-phenylphenol (R=C₁ to C₁₀). The products, which in most instances are calix[6]arenes, calix[7]arenes, and calix[8]arenes, are cited as absorbents for heavy metal ions. For example, it is claimed that a mixture made from the calixarene and a solution containing copper ions shows an increased rate of copper deposition for plating solutions, useful for forming copper coatings on plastics, ceramics, and printed circuit boards. The procedures described in these patents for preparing the calixarenes appear to be essentially identical with those previously described

---

[9] Hitachi Chemical Co., Ltd., Jpn. Kokai Tokkyo Koho JP 59,104,331, **1984** (*Chem. Abstr.*, 101: 191410); *ibid.*, JP 59,12,913, **1984** (*Chem. Abstr.*, 101: 24477); *ibid.*, JP 59,12915, **1984** (*Chem. Abstr.*, 101: 8163); *ibid.*, 60,202,113, **1984** (*Chem. Abstr.*, 104: 207904); *ibid.*, JP 61,106,775, **1986** (*Chem. Abstr.*, 105: 231158k).

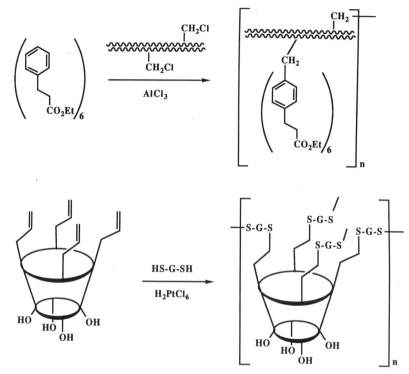

Figure 7.2 *Polymeric calixarenes* via *'upper rim' attachment*[8]

in the literature. Two more patents[10] were issued to this same company in 1987 describing polymeric calixarenes that are stated to have good metal absorptivity and heat resistance. They were derived from *p*-phenylcalixarenes by acylation with RCOCl (R is $CH=CH_2$, $C(Me)=CH_2$, or $(CH_2)_mCH=CH_2$) which introduces unsaturated acyl groups into the 4'-position of the biphenyl ring. Polymerization in the presence of methyl methacrylate was induced with AIBN to give a copolymer with a softening point at 180—185 °C.

Electrodes selective for Na^+ (based on calixarene esters and amides) and for Na^+ and Cs^+ (based on *p*-alkylcalixarene acetates) have been described by Kimura and coworkers[11] and by Diamond.[12] The electrodes are stated to work well as potentiometric sensors, to show good selectivity for the primary ion, to have virtually no response to divalent cations, and to be stable over a

[10] Y. Nakamoto, S. Ishida, and Y. Yoshimura (Hitachi Chemical Co., Ltd.), Jpn. Kokai Tokkyo Koho JP 62,96,505, **1987** (*Chem. Abstr.*, 108: 56797f); *idem., ibid.*, 62,96,440, **1987** (*Chem. Abstr.*, 108: 6959q).

[11] K. Kimura, M. Matsuo, and T. Shono, *Chem. Lett.*, **1988**, 615.

[12] D. Diamond, *Anal. Chem. Symp. Ser.*, **1986**, 25 (*Chem. Abstr.*, 106: 148452b); *idem.* in *Electrochemistry, Sensors, and Analysis*, Proc. Intl. Conf., Dublin, June 10—12, **1986**, Vol. 25, p. 155.

wide pH range. A 'lower rim' bridged calixarene, the calixspherand (see **10** in Figure 4.16) of Reinhoudt, Ungaro, and coworkers[13] has been synthesized as part of a program concerned with ^{81}Rb$^+$ imaging.

A pair of interesting patents have been issued to Shinkai and his co-workers[14] for the use of calixarenes as uranophiles, as described in Chapter 6. The world's oceans contain a total of about 3 billion tons of uranium, present as UO_2^{2+} in association with carbonate. Although this represents an enormous quantity of material its concentration is only about 3 parts per billion, an amount that corresponds to less than 1 mg in a large sized back yard swimming pool. Also, the UO_2^{2+} is accompanied by numerous other cations, most of them present in much larger concentration. Thus, the extraction of uranium from sea water poses a tantalizing challenge which has been addressed by a number of chemists during the past decade. If the 'greenhouse effect' proves to be responsible for adverse global changes in climate it may force greater attention to nuclear energy, and the recovery of uranium from sea water will become a more pressing problem in the future than it is at the present time. One of the first reports of UO_2^{2+} complexation came from the laboratories of Donald Cram[15] in 1976 where crown ether chemistry was coming to fruition. The compound they tested was found to have a K_{assoc} of 10^{11} M^{-1} for this cation. Much of the recent work in this field has come from Japan under the leadership of the late Iwao Tabushi who found that the macrocyclic triamine **2** is particularly effective,[16] having a K_{assoc} of $10^{20.7}$ M^{-1} for UO_2^{2+}. This material was put to an actual sea test[17] by placing it 10 m below the surface of the ocean in the strong Kuorshio Current off the coast of Mikura Island in Japan where it absorbed 50 μg of uranium per gram of resin

2

[13] D. N. Reinhoudt, P. J. Dikjkstra, P. J. A. in't Veld, K. E. Bugge, S. Harkema, R. Ungaro, and E. Ghidini, *J. Am. Chem. Soc.*, **1987**, *109*, 4761.

[14] S. Shinkai, O. Manabe, Y. Kondo and T. Yamamoto (Kanebo Ltd.), Jpn. Kokai Tokkyo Koho, JP 62,136,242, **1987** (*Chem. Abstr.*, 108: 64410q); Y. Kondo, T. Yamamoto, O. Manabe, and S. Shinkai (Kanebo Ltd.), Jpn. Kokai Tokkyo Koho, JP 62,210,055, **1987** (*Chem. Abstr.*, 108: 116380b).

[15] A. H. Alberts and D. J. Cram, *J. Chem. Soc., Chem. Commun.*, **1976**, 958.

[16] I. Tabushi, Y. Kobuke, and A. Yoshizawa, *J. Am. Chem. Soc.*, **1984**, *106*, 2481.

[17] I. Tabushi, Y. Kobuke, N. Nakayama, T. Aoki, and A. Yoshizawa, *I & EC Prod. Res. Dev.*, **1984**, *23*, 445.

per day. Even more specific than **2** for UO_2^{2+}, however, are the *p*-sulfonato-calix[6]arenes. A polymer-bound analog of this material is described in a Shinaki patent which states that the hexakis(carbethoxymethyl) ether of *p*-sulfonatocalix[6]arene is 'partially nitrated, aminated, and fixed on cross-linked chloromethylated polystyrene'. This resin is stated to absorb 108 μg of uranium from sea water per 0.1 g of resin in 7 days at a flow rate of 30 ml/min. Another polymer-bound analog has also been reported by Shinkai *et al.*[18] who treated *p*-(chlorosulfonyl)calix[6]arene with polyethyleneimine and obtained a gel-like product which contained one calixarene unit for every 15 ethyleneimine units. It showed the same binding power and selectivity for UO_2^{2+} as the parent calixarene, *p*-sulfonatocalix[6]arene.

7.2 Calixarenes in Industry

The increasing number of patents describing calixarenes attests to the growing interest in the utilitarian aspects of these compounds. Though much of this interest is shrouded by the veil of industrial secrecy, the disclosures in patents provide a few clues as to where these interests lie and in what directions they may be heading.

The extent to which calixarene chemistry plays a part in the manufacture of Bakelite remains uncertain. The possibility that cyclic structures, *i.e.* calixarene-like structures, are present in Bakelite had its inception in a suggestion by Raschig[19] in 1912. But, as Baekeland pointed out at the time[20] 'we must not forget that one hypothesis is about as easy to propose as another so long as we are unable to use any of the methods of determining molecular size and molecular construction'. Of course, many techniques for determining molecular size and construction have become available since the time of Raschig and Baekeland, and these have provided considerable insight into the structure of phenol–formaldehyde resins. But, the complete structure remains sufficiently unclear to still attract the attention of chemists using the newest techniques such as crossed polarization 'magic angle' spinning spectroscopy (CPMAS) which Maciel has applied to phenol–formaldehyde materials.[21]

The Petrolite demulsifier discussed in Chapter 1 is an industrial phenol–formaldehyde material in which calixarenes unquestionably do play a role. The oxyalkylates of *p*-alkylphenol–formaldehyde condensates from which the resin is prepared by what has come to be known as the 'Standard Petrolite Procedure' are mixtures containing a large component of calix[8]arene along with a certain amount of the calix[6]arene. The problem of insoluble sludges (see Chapter 1) has been circumvented either by blending the calixarene-derived oxyalkylate with oxyalkylates prepared from linear oligomers or by

[18] S. Shinkai, H. Kawaguchi, and O. Manabe, *J. Polymer Sci., Part C, Polymer Lett.*, **1988**, *26*, 391.
[19] F. Raschig, *Z. Angew. Chem.*, **1912**, *25*, 1939.
[20] L. H. Baekland, *J. Ind. Eng. Chem. (Industry)*, **1913**, *5*, 506.
[21] G. E. Maciel and I.-S. Chuang, *Macromolecules*, **1984**, *17*, 1081; G. R. Hatfield and G. E. Maciel, *ibid.*, **1987**, *20*, 608.

using a phenol with a larger *p*-alkyl group, *i.e.* octyl or nonyl. Tests at Petrolite have shown that the oxyalkylates derived from the cyclic oligomers are superior demulsifiers to those derived from linear oligomers. The reasons for this are quite obscure, and it can only be surmised that demulsifying action is dependent on the aggregation characteristics of the calixarene portion of the oxyalkylate and on the manner in which this portion of the molecule interacts with the lipophilic components of crude oil.

The industrial application of the interaction of calixarenes with inorganic cations has already been discussed in the case of the Loctite and Hitachi patents. Perhaps related to this is the patent issued to Pastor and Odorisio[22] which claims antioxidant and heat stabilizing properties (for plastics) for acylated calixarenes. A patent reminiscent of the chemistry described by Muthukrishnan and Gutsche[23] in 1979 has been issued to a Japanese group[24] who prepared a penta-*p*-nitrophenyl ether of *p-tert*-butylcalix[8]arene, stating it to be useful as an artificial enzyme and for heat-resistant polyimides. The extraordinary ease with which calixarenes such as those described above can be synthesized makes them immediately attractive candidates for commercial application, and numerous industries throughout the world are cognizant of their possibilities. There is no doubt that this interest goes well beyond that reflected in the current patent literature, and new uses are likely to be revealed during the next several years.

7.3   Calixarenes as Complexers of Neutral Compounds

Selective complexation has been demonstrated by crystallizing *p-tert*-butylcalix[4]arene from 50:50 mixtures of pairs of guest molecules such as benzene and *p*-xylene,[25] one of the host–guest combinations precipitating preferentially. Anisole and *p*-xylene, for example, are complexed in preference to most other simple aromatic hydrocarbons. Attempts to exploit this selectivity by using calixarenes as immobile phases on chromatographic columns have been only marginally successful.[26] Calixarenes as the stationary phases for column chromatography have been shown by the Parma group[27] to be effective in the separation of alcohols, chlorinated hydrocarbons, and aromatic compounds.

[22] S. D. Pastor and P. Odorisio, U.S. Patent 4,617,336, **1986** (*Chem. Abstr.*, 106: 6038x).

[23] R. Muthukrishnan and C. D. Gutsche, *J. Org. Chem.*, **1979**, *44*, 3962.

[24] H. Taniguchi, E. Nomura, and R. Maeda, Jpn. Kokai Tokkyo Koho JP 62,233,156, **1987** (*Chem. Abstr.*, 109: 55418s).

[25] G. D. Andreetti, A. Mangia, A. Pochini, and R. Ungaro, Abstracts of the 2nd International Symposium on Clathrate Compounds and Molecular Inclusion Phenomena,' Parma Italy, **1982**, p. 42.

[26] E. Smolkova-Keulemansova and L. Felti, ref. 25, p. 45.

[27] A. Mangia, A. Pochini, R. Ungaro, and G. D. Andreetti, *Anal. Lett.*, **1983**, *16*, 1027.

7.4 Calixarenes as Catalysts

In contrast to the cyclodextrins, where a number of suitably modified members of this class of compounds have been shown to have catalytic properties, only one good example of a well-defined calixarene-catalyzed process has been published. Shinkai and coworkers[28, 29] have used a reaction investigated some years ago by Metzler[30] and Chaykin[31] involving the acid-catalyzed addition of water to 1-benzyl-1,4-dihydronicotinamide (**3**) to form 1-benzyl-6-hydroxy-1,4,5,6-tetrahydronicotinamide (**4**), as illustrated in Figure 7.3. The reaction is carried out at 30 °C in an aqueous solution

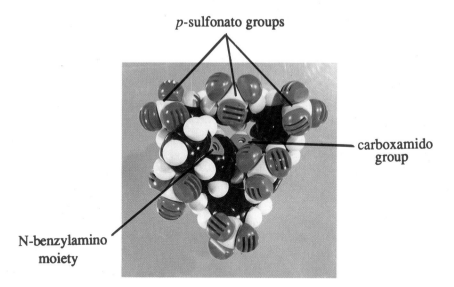

3   **4**

Figure 7.3 *Acid-catalyzed addition of water to 1-benzyl-1,4-dihydronicotinamide*

p-sulfonato groups

carboxamido group

N-benzylamino moiety

CPK Model of *p*-sulfonatocalix[6]arene with *N*-benzyl-1,4-dihydronicotinamide in the cavity

[28] S. Shinkai, H. Koreishi, S. Mori, T. Sone, and O. Manabe, *Chem. Lett.*, **1985**, 1033.
[29] S. Shinkai, S. Mori, H. Koreishi, T. Tsubaki, and O. Manabe, *J. Am. Chem. Soc.*, **1986**, *108*, 2409.
[30] C. J. Johnston, J. L. Gardner, C. H. Suelter, and D. E. Metzler, *Biochemistry*, **1963**, *2*, 689.
[31] C. S. Y. Kim and S. Chaykin, *Biochemistry*, **1968**, 7, 2339.

buffered, in most cases, at pH 6.30 and is conveniently followed by measuring the shift in the absorption band from 340—360 nm (1,4-dihydro compound) to *ca.* 290 nm (tetrahydro compound). Plots of the rate of the reaction *vs.* the concentration of the calixarene (**5a—e**) show saturation kinetics, suggesting that the reaction scheme is one that is typical for a process involving the formation of a complex, *viz.*

5a   R = H

5b   R = CH$_2$CO$_2$H

5c   R = Me

5d   R = *n*-hexyl

5e   R = *n*-dodecyl

Analysis of the kinetic results by standard methods produced the numbers shown in Table 7.1 where, in addition to the calixarenes **5a—d**, the comparison compounds sodium dodecyl sulfate (SDS), *p*-hydroxybenzenesulfonic acid (PBS), and *p*-(carboxymethoxy)benzenesulfonic acid (CPBS) are also included. Inspection of these data reveals that calixarenes having a protic 'lower rim' (*e.g.* **5a** and **5b**) are better catalysts than those lacking this feature (*e.g.* **5c** and **5d**) and that the cyclic oligomers are more effective than their acyclic analogs, the latter showing simple second-order reaction kinetics.

The hydration reaction shown in Figure 7.3 is enzyme-catalyzed by glyceraldehyde 3-phosphate dehydrogenase, and it has been suggested[32] that cationic stabilization as well as proton donation are requisite catalytic features in this reaction, as illustrated in Figure 7.4. Thus, it is postulated[29] that *p*-sulfonatocalixarene **5a** and its carboxymethyl ether **5b** behave in a com-

Table 7.1 *Rates of conversion of 1-benzyl-1,4-dihydronicotinamide (3) into 1-benzyl-6-hydroxy-1,4,5,6-tetrahydronicotinamide (4) in the presence of calixarenes (5)[29]*

Catalyst	pH	$k_1/10^{-4}$	$k_c/10^{-4}\,s^{-1}$	K/M^{-1}
5a	6.30	46.2	131	564
5b	6.30	25.5	47.5	1340
5b	4.00	328	668	1020
5c	6.30	0.36	3.0	287
5d	6.30	3.31	6.2	2160
SDS	6.30	1.13		
PBS	6.30	0.038		
CPBS	4.00	0.77		

[32] D. Moras, K. W. Olsen, M. N. Sabesan, M. Buehner, G. C. Ford, and M. G. Rossman, *J. Biol. Chem.*, **1975**, *250*, 9137; G. Branlant, D. Tritsch, B. Eiler, L. Wallen, and J.-F. Bielmann, *Eur. J. Biochem.*, **1982**, *129*, 437.

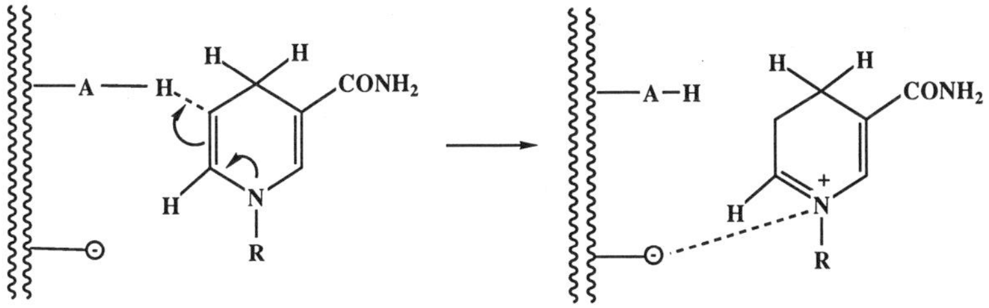

Figure 7.4 *Mechanism of catalysis of hydration of 1,4-dihydronicotinamides*[29]

parable fashion, providing anionic sites at the 'upper rim' region (carrying the sulfonate groups) and proton donating sites at the 'lower rim' (carrying the phenolic or carboxyl groups). That the guest molecule **3** is probably inside the cavity of the calixarene is indicated by fluorescence measurements. In the presence of SDS, which is a micelle-forming compound that provides a hydrophobic environment for an occluded guest molecule, the fluorescence intensity of **3** decreases. This is also observed in the presence of calixarene **5d** (for which a K_{assoc} of 2860 M^{-1} is calculated from the spectral data). With calixarenes **5a** and **5b**, however, the fluorescence intensity *increases*, suggesting that these compounds provide a microenvironment that is *more* polar than water, a conclusion that finds additional support from the complexation experiments with Phenol Blue discussed in Chapter 6.

Potentiometric titrations show that compound **5a** has five undissociated protons at pH 6.30, while compound **5b** has only two. This may account for the larger k_c for **5a**, and support for this idea is given by the fact that **5b** becomes a much better catalyst at pH 4.0 than at pH 6.30. The larger K_{assoc} for **5b** compared with **5a** is attributed to the deeper cavity presented by **5b**, although the fact that the K_{assoc} for **5a** is larger than that of **5c** indicates that other forces such as hydrogen bonding must also play a role.

Calixarene catalysis of the hydration of 1-benzyl-1,4-dihydronicotinamide has also been studied by Gutsche and Alam[33] who tested the entire series of *p*-(carboxyethyl)calixarenes (**6**) from the cyclic tetramer through the cyclic octamer. Although the cyclic tetramer is ineffective, all of the other members of this series produce measurable catalysis, the maximum occurring with the cyclic hexamer as indicated by the data in Table 7.2. Comparison of the values in Table 7.2 with those in Table 7.1 show that *p*-sulfonatocalix[6]arene is about four times more effective as a catalyst than *p*-(carboxyethyl)calix[6]arene, although the latter binds the substrate somewhat more tightly. In fact, in the *p*-)carboxyethyl)calixarene series, there appears to be an inverse relationship between the K_{assoc} values and the k_c values. That tightness of binding is not the *sine qua non* for catalysis is an observation that has been made in many other systems as well. The superiority of *p*-sulfonatocalix[6]arene (**5a**) over *p*-(carboxyethyl)calix[6]arene (**6**, *n* = 6) may arise

[33] C. D. Gutsche and I. Alam, *Tetrahedron*, **1988**, *44*, 4689.

Table 7.2 *Rates of conversion of 1-benzyl-1,4-dihydronicotinamide (3) into 1-benzyl-6-hydroxy-1,4,5,6-tetrahydronicotinamide (4) in the presence of calixarenes (6)[33]*

	n	$k_1/10^{-4}\,s^{-1}$	$k_c/10^{-4}\,s^{-1}$	K_{assoc}/M^{-1}
	5	11	13	3800
	6	17	31.9	880
	7	4.4	6.0	5000
	8	4.6	6.6	3700
	4-hydroxyphenyl-propionic acid	0.2		

6

from the greater concentration of negative charge at the 'upper rim' of the calix, the six sulfonato-groups being held more rigidly in place than the more flexible carboxyethyl moieties.

A catalyst, defined as a substance that alters the rate of a reaction without itself undergoing any net change, is generally perceived in terms of rate enhancement, *i.e.* positive catalysis. Rate reduction or negative catalysis, however, can also be a manifestation of the catalytic phenomenon, and an example of this is provided by reactions of aryldiazonium salts in the presence of the *p*-sulfonatocalixarenes. Shinkai and coworkers[34,35] showed that not only do these two species form complexes with one another, as discussed in Chapter 6, but the complexed guest reacts at a different rate than the uncomplexed guest. Thus, the rate ratios k/k_0 that are observed for benzenediazonium and *p*-hexylbenzenediazonium cations undergoing denitrification to form phenols shown in Table 7.3, are all less than 1.0. Calixarene **5e**, for example, shows a five-fold rate reduction. In a similar vein, the coupling reaction of *p*-chlorobenzenediazonium cation with *p*-dimethylaminoaniline and 3-hydroxy-2,7-naphthalenedisulfonic acid show, with one

Table 7.3 *Rate ratios for denitrification and coupling reactions of aryldiazonium cations in the presence of sulfonatocalix[6]arenes[35]*

Added reagent	R—ArN$_2^+$ → R—ArOH		Coupling reagent	
	R = H	R = Me(CH$_2$)$_5$	Me$_2$N—⟨⟩—NH$_2$	HO$_3$S⟨⟩SO$_3$H / OH
	k/k_c	k/k_c	k/k_c	k/k_c
none	1.00	1.00	1.00	1.00
5a	0.91	0.79	1.53	0.6
5d	0.31	0.23	0.091	0.024
5e	0.20	0.23	0.077	0.016

[34] S. Shinkai, S. Mori, K. Arabi, and O. Manabe, *Bull. Chem. Soc. Jpn.*, **1987**, *60*, 3679.
[35] S. Shinkai, S. Mori, T. Arimura, and O. Manabe, *J. Chem. Soc., Chem. Commun.*, **1987**, 238.

exception, reductions in rate. Compounds **5d** and **5e** are particularly effective in this respect.

Since the complexation constants for aryldiazonium salts interacting with calixarenes are considerably greater for **5d** and **5e** than for **5a**, it was concluded that hydrophobic as well as electrostatic forces must be responsible for the rate reductions in the denitrification reaction. The data for the coupling reactions are interesting in showing a slight rate enhancement with **5a** in one case but quite large rate reductions with **5d** and **5e**, two variables being invoked to interpret these results, *viz.* cation stabilization (leading to rate reduction) and reactant concentration in the calixarene cavity. The consequences of the latter, however, are uncertain in the absence of precise knowledge of the nature of the calixarene cavity.

7.5 Calixarenes as Biomimics

For many researchers, including this author, a major interest in the calixarenes resides in their potential for providing molecules capable of mimicking various aspects of natural systems. To date, however, this interest remains a promise awaiting fulfillment. Only the system described above involving the hydration of a dihydronicotinamide qualifies as a *bona fide* example of calixarene catalysis, and this area of calixarene chemistry continues to offer the greatest challenge to the ingenuity of its explorers. The following section presents a pair of speculative proposals for calixarene-derived biomimics to whet the reader's appetite to design, construct, and test other such systems in the quest for interesting and useful catalysts.

7.5.1 Heme Mimics

The search for simple, well-defined systems that can reversibly bind oxygen and act as heme mimics has occupied the attention of a number of chemists for the last two decades, and some remarkable advances have been made. Most of the systems that have worked employ a porphyrin ring in one way or another, thus quite closely resembling the natural systems. Further removed from the structure of the porphyrins are the lacunar compounds invented by Daryl Busch and his colleagues at Ohio State University.[36] Still further removed is Cram's cavitand[37] (see **14** in Figure 4.25) which shows some affinity for oxygen. In the hope of developing another non-porphyrin heme mimic, Gutsche and Nam[38] investigated the metal complexation behavior of *p*-(2-aminoethyl)calix[4]arene, synthesized by the *p*-quinonemethide procedure described in Chapter 5. The rationale for this work lies in the ability

---

[36] D. H. Busch and C. Cairns, 'Ligands Designed for Inclusion Complexes: From Template Reactions for Macrocyclic Ligand Synthesis to Superstructured Ligands for Dioxygen and Substrate Binding' in 'Progress in Macrocyclic Chemistry', Vol. 3, 'Synthesis of Macrocycles', ed. R. M. Izatt and J. J. Christensen, John Wiley, New York, **1987**, p. 1.

[37] D. J. Cram, K. D. Stewart, I. Goldberg, and K. N. Trueblood, *J. Am. Chem. Soc.*, **1985**, *107*, 2574.

[38] C. D. Gutsche and K. C. Nam, *J. Am. Chem. Soc.*, **1988**, *110*, 6153.

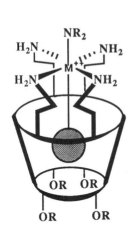

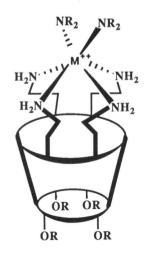

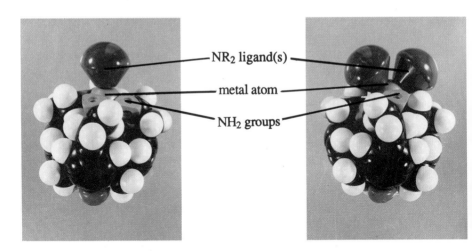

Structure A - one *endo*-calix and one
exo-calix site on metal

Structure B - two *exo*-calix sites
on metal

Figure 7.5 *CPK Models and stylized representations of two configurations of the hexacoordinate octahedral complexes of* **7a**

of the four aminoethyl groups to align their ligands in a square planar fashion around a cation such as Fe^{2+}, leaving a fifth site on the metal for coordination with an external Lewis base and a sixth site oriented inward, accessible only to molecules small enough to traverse the annulus at the 'lower rim' of the calixarene, as illustrated in Figure 7.5. The complexation studies were carried out with the *p*-bromobenzenesulfonate ester **7b** to insure a rigid cone conformation and also to obviate any involvement of the phenolic hydroxyl groups in the interaction with the metal ion. Although **7b** engages a variety of metals to form complexes whose spectral characteristics give some insight into their geometry, none of the complexes show any affinity for oxygen. One possible explanation for this failure comes from inspection of CPK models which indicate that octahedral complexation can occur not only in the desired fashion depicted by structure A in Figure 7.5 but also in a fashion that orients *both* the fifth and sixth coordination sites external to the calix, as depicted by structure B in Figure 7.5. A tetraamine that allows metal complexation to produce only the A-type structure and that also provides 'side portals' for the entrance and exit of small molecules is the *p*-imidazolylmethylcalix[4]arene (**8**), readily made *via* the *p*-quinonemethide method. Its properties as a complexing agent remain to be explored.

7a   R = H

7b   R = SO₂C₆H₄Br

8

7.5.2   Aldolase Mimics

One of the many interesting reactions in the carbon cycle of photosynthesis is the condensation of dihydroxyacetone phosphate (DHAP) and glyceraldehyde phosphate (GAP) to form fructose-1,6-diphosphate, the reverse reaction occurring when carbohydrates are metabolized. Aldolase, the enzyme that catalyzes this reaction, has an active site that contains (a) a positively charged region, which forms an electrostatic bond with the phosphate oxygens of DHAP, (b) a primary amine (lysine-57 residue) which interacts with the carbonyl group to form a protonated Schiff base, anchoring the substrate to the enzyme and increasing the acidity of the α-hydrogen, and (c) a basic moiety (probably an imidazole of a histidine residue) which displaces an α-hydrogen and generates the reactive intermediate that effects a nucleophilic addition to the carbonyl group of GAP to form the new carbon–carbon bond.

In contrast to heme mimics where numerous systems have been devised, few aldolase mimics have been proposed and tested.[39,40] In response to this dearth of candidates a suggestion was made in an earlier review[41] for an aldolase mimic based on a calixarene. The proposed compound (**9**) contains a pair of diphenyl moieties to increase the depth of the cavity, a bridge between the diphenyl moieties to reduce the conformational flexibility, carboxymethyl groups affixed to the phenolic oxygens to establish a cone or partial cone conformation and to increase the water solubility of the compound, and carboxyl and amino groups on the 'upper rim' and 'middle rim' of the cavity to perform the chemistry of the process. The rationale for proposing **9** to serve as an aldolase mimic assumes that a metal atom coordinated

9

with DHAP will be attracted to the cavity through electrostatic interaction with the oxygens at the 'lower rim' of the calixarene, that a primary amino group will engage the carbonyl group of DHAP in Schiff base formation, and that one of the carboxylate groups at the 'upper rim' of the calix will act as the basic moiety to remove an α-H and generate the enamine in the manner depicted in Figure 7.6. Condensation of the enamine with GAP can then take place along a pathway unhindered by the calixarene to form the product and allow the process to reverse and start afresh with another pair of substrate molecules.

7.6 Calixarenes as Physiological Compounds

Phenolic compounds are well known for possessing physiological properties. Good examples are the urushiols which are long chain alkyl-substituted

[39] C. D. Gutsche, D. Redmore, R. S. Buriks, K. Nowotny, H. Grassner, and C. W. Armbruster, *J. Am. Chem. Soc.*, **1967**, *89*, 1235.
[40] G. A. Gettys and C. D. Gutsche, *Bioorg. Chem.*, **1978**, *7*, 141.
[41] C. D. Gutsche, *Acc. Chem. Res.*, **1983**, *16*, 161.

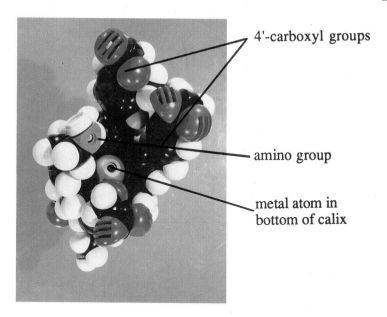

4'-carboxyl groups

amino group

metal atom in
bottom of calix

CPK Model of an aldolase mimic

catechols present as the active vesicant principle of poison ivy. Qualitatively comparable dermatitic responses have been noted for *p-tert*-butylphenol–formaldehyde resins, particularly, the linear tetramer.[42] *p-tert*-Butylcalix[4]arene and *p-tert*-butylcalix[8]arene give negative responses in the Ames test for mutagenicity, but whether this truly reflects an innocuous character or just their great insolubility is not certain.

An extensive medical and biochemical literature has arisen during the last three decades concerning the oxyalkyl derivatives of simple phenols as well as phenol–formaldehyde condensation products. It was the quest of such compounds, in fact, that led Cornforth to reinvestigate the Zinke reaction in the 1950's and to lay some of the groundwork for subsequent developments in calixarene chemistry. In Cornforth's initial work[43] he oxyethylated calixarenes by treating them with ethylene oxide which yielded a product called a 'macrocyclon'. In a follow-up investigation some years later,[44] a more carefully controlled set of experiments was carried out using the high melting compound from the condensation of *p*-(1,1,3,3-tetramethylbutyl)phenol and formaldehyde (Cornforth's HOC compound, subsequently shown to be the cyclic octamer). Tests of these compounds for tuberculostatic activity led Cornforth and coworkers to conclude that the lipophilic–hydrophilic balance of the molecule may be the most critical factor but that resistance to chemical

[42] H. Schubert and G. Agatha, *Dermatosen in Beruf and Umwelt*, **1979**, *27*, 49.
[43] J. W. Cornforth, P. D'Arcy Hart, G. A. Nicholls, R. J. W. Rees, and J. A. Stock, *Br. J. Pharmacol.*, **1955**, *10*, 73.
[44] J. W. Cornforth, E. D. Morgan, K. T. Potts, and R. J. W. Rees, *Tetrahedron*, **1973**, *29*, 1659.

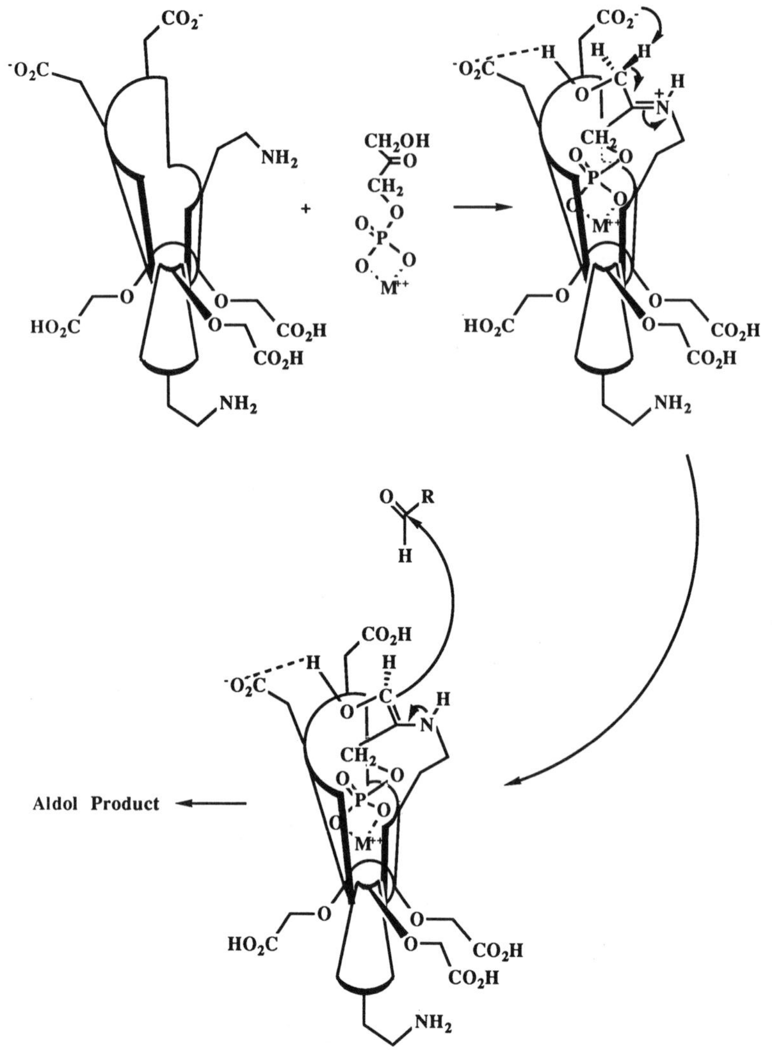

Figure 7.6 *Proposed mechanism of action of compound* **9** *acting as an aldolase mimic*[41]

breakdown *in vivo* is also important if activity is to be shown against the slowly developing experimental tuberculosis. Macrocyclon has been tested by Delville and Jacques[45] in the therapy of parasitic diseases, and Hart *et al.* have used it to induce fusion of erythrocytes.[46] The most recent investigation in which macrocyclon and related compounds have been used is that of Jain

[45] J. Delville and P. J. Jacques. *Biochem. Soc. Trans.*, **1978**, *6*, 395.
[46] C. A. Hart, Q. F. Ahkong, D. Fisher, T. Hallinan, S. J. Quirk, and J. A. Lucy, *Biochem. Soc. Trans.*, **1978**, *3*, 733.

and Jahagirdar[47] who studied the effects of phospholipase A_2. Obtaining results reminiscent of those of earlier workers investigating other physiological responses, they found that a calixarene (Cornforth's HOC compound) carrying relatively short polyoxyethylene chains on the oxygens (*ca.* 12.5 units on each ether oxygen) inhibited the action of phospholipase A_2 while those having longer polyoxyethylene chains (*ca.* 60 units on each ether oxygen) stimulated its action. It would be interesting to determine whether the physiological activity of these compounds is dependent primarily on the extent of oxyalkylation or whether it is also dependent on the size of the calixarene ring and the substituents in the *p*-positions.

As a concluding comment on the possible physiological attributes of calixarenes we quote the visionary statements of Donald Cram[48] who says of the carcerands that 'Large metabolizable molecular cells might be used in drug or agricultural chemical delivery systems or in systems in which very slow release of compounds chemically shielded from their environment are needed. Metabolism-resistant carcerands containing appropriate radiation-

Crown Ether   **Cyclodextrin**

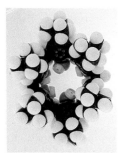

Calixarene

CPK Models of 18-crown-6, α-cyclodextrin, and *p-tert*-butylcalix[6]arene

[47] M. K. Jain and D. V. Jahagirdar, *Biochem. J.*, **1985**, *227*, 789.
[48] D. J. Cram, S. Karbach, Y. H. Kim, L. Baczynskyj, K. Marti, R. M. Sampson, and G. W. Kalleymeyn, *J. Am. Chem. Soc.*, **1988**, *110*, 2554.

emitting metal ions or atoms might be attached to immunoproteins that seek out cancer cells. Others might provide means for organ imaging. The shell of the host should inhibit deposition of heavy metal salts in bonds. The unusual thermal stability of the carcerand and its molecular shape suggest possible uses as molecular ball bearings'.

7.7 Concluding Remarks

The calixarenes are assuming a place in supramolecular chemistry alongside the cyclodextrins and crown ethers, to which they have been compared in various discussions in this book. Although the calixarenes have developed much less rapidly than their companions, interest in their potential appears to be increasing and attention to their chemistry is escalating. Their particular charm is their easy accessibility, acid-catalyzed condensations providing high yields of the resorcinol-derived calix[4]arenes and base-induced condensations providing high yields of phenol-derived calixarenes in a variety of cavity sizes. The means are clearly at hand for making molecular baskets that can be constructed by design to perform myriad tasks. Great progress can be anticipated in the years ahead in this field of endeavor to which Ulysses' reflections are apropos:

> *'Yet all experience is an arch*
> *Wherethrough gleams that untravelled world,*
> *Whose margin fades*
> *For ever and for ever when I move'*

Alfred Lord Tennyson, *Ulysses*

Indexes

Index A: Subject Index*

*For specific compounds see Index B (Phenol-Derived Calixarenes and Index C (Resorcinol-Derived Calixarenes).

Böhmer, Volker, biography and picture,
39–40
Bridged calixarenes, upper rim – see
Arrichoarenes
Bromination, of calixarene ethers, 138,
146
Bromoresorcinol – see Resorcinol,
bromo
Bruice, Thomas, 173
Buriks, Rudolf, 17–18
Busch, Daryl, 197

Calix, in calixarene nomenclature, 20
Calix crater, 20, 21
Calix[4]arenes – also see Calixarene
complexes, 149–152
conformations, 88, 90–97, 101
functionalized ether derivatives,
131–134
link between one-step and multi-step
synthesis, 73
mass spectra, 84
mechanism of formation, 52, 53, 55
methylene resonances, Table, 108
nitration of *p-tert*-butyl, 136
O-acetylation, 138
O-benzoylation, 138
one-step synthesis, 28–32
p-tert-butyl, treatment with nitric
acid, 136
stabilization of cone conformation
by trimethylanilinium, 176
stepwise synthesis, 38–45
tetra-trimethylsilyl ethers, 131
upper rim bridged – see Upper rim,
bridged
Calix[4]quinone, 147
Calix[4]resorcinarenes – also see
resorcinol-derived calixarenes,
32–34, 72–73, 101–105, 124,
145–146
complexes, 156–158, 168–170,
177–178
configuration, 72, 89–90, 101–105
conformational interconversion,
101–104
functionalization, 145–146
inversion barrier, 103, 104
mass spectra, 83

pseudorotation, 103–104
solid state conformations, 90
upper rim bridged, NMR spectrum,
123–124
X-ray crystallography, 72
Calix[5]arenes – also see Calixarene
complexes, 149, 152
conformation, 89, 98
one-step synthesis, 45, 85
stepwise synthesis, 39–70
X-ray structure, 69
Calix[6]arenes – also see Calixarene
complexes, 149
conformational interconversion,
111–113
conformations, 89, 97–98, 101
functionalized ether derivatives,
131–134
hexa-trimethylsilyl ether, 131
mechanism of formation, 54
stepwise convergent synthesis, 47
X-ray structure, 69
Calix[6]arene oxyanions – see
Oxyanions, calix[6]arenes, 124
Calix[7]arenes – also see Calixarene
complexes, 149
conformational inversion, 99
conformation, 89, 99, 101
stepwise synthesis, 38
Calix[8]arenes – also see Calixarene
complexes, 149, 152
conformational interconversion, 96
conformations, 89, 94–96, 101
functionalized ether derivatives,
112, 132–134
mass spectra, 83
mechanism of formation, 52–54
obligatory intermediate in
calix[4]arene formation, 56
octa-trimethylsilyl ether, 131
octaacetoxy, 131
octamethyl ether, reaction with
Me₃Al, 122
pinched conformation and complexation, 175
stepwise synthesis, 47
structure proof, 69
X-ray structure, 69
Calix[9]arenes – also see Calixarene,
85, 99

Index B: Phenol-Derived Calixarenes

$$\left(\begin{array}{c} R^1 \\ \\ CH_2 \\ OR^2 \end{array} \right)_n$$

R^1	R^2		n	Page

Calixarene Ethers

R¹	R²	n	Page
t-Butyl	CH$_2$CO$_2$Me	4	132–134, 161–163
t-Butyl	CH$_2$CO$_2$Me	6	132–134, 161–163
t-Butyl	CH$_2$CO$_2$Me	8	132–134, 161–163
t-Butyl	SCONMe$_2$	4	107
t-Butyl	Methyl, Me$_3$Al	4	120–122
t-Butyl	Methyl, Me$_3$Al	6	120–122
t-Butyl	Methyl, Me$_3$Al	8	120–122
t-Butyl	Titanium	4	116
t-Butyl	Iron	4	116, 117, 118
t-Butyl	Chromium	4	116, 117, 118
t-Butyl	Titanium	6	118, 154
t-Butyl	Europium	8	118, 119
t-Butyl	four Methyl, Titanium	6	120, 154
t-Octyl	Methoxyethyl	8	153
Allyl	Methyl	6	69–70, 111
Allyl	two Benzyl	4	130

p-Substituents Containing Hetero Atoms

R¹	R²	n	Page
CH$_2$Cl	Methyl	6	144
CH$_2$Cl	Methyl	8	144
CN	Methyl	4	138
CO$_2$H	Methyl	4	138
CO$_2$H	Methyl	6	138
CO$_2$H	Methyl	8	138
Acetyl	H	4	137
Acetyl	Methyl	6	76, 137
Acetyl	Methyl	8	138
Benzoyl	Methyl	4	138
p-Methoxybenzoyl	Methyl	4	138–139
p-Hydroxybenzoyl	H	4	139
SO$_3$H	Methyl	6	175, 190, 194
SO$_3$H	EtC(Me)CH$_2$	6	48, 175, 180
SO$_3$H	*n*-Hexyl	6	175, 179, 194–195, 196–197
SO$_3$H	*n*-Dodecyl	6	175, 179–180, 194–195, 196–197
SO$_3$H	CH$_2$CO$_2$H	4	174
SO$_3$H	CH$_2$CO$_2$H	5	174
SO$_3$H	CH$_2$CO$_2$H	6	137, 174, 194–195
SO$_3$H	CH$_2$CO$_2$Et	6	191

R^1	R^2	n	*Page*

Calixarene Esters

p-H Substituents

H	Acetyl	4	76
H	Benzoyl	4	109
H	three Benzoyl	4	140
H	three Benzoyl, Allyl	4	140

p-Alkyl and Alkenyl Substituents

t-Butyl	Tosyl	4	108
t-Butyl	Acetyl	4	69–70, 74, 106, 107, 128, 150
t-Butyl	three Acetyl	4	130
t-Butyl	Acetyl	6	112, 130
t-Butyl	Acetyl	8	69–70, 130, 153
t-Butyl	Benzoyl	4	109
t-Butyl	*p*-Y-Benzoyl (Y=OMe, O-*t*-Bu, Me, Br, CF_3, CN, NO_2)	4	129
two *t*-Butyl	two 3,5-Dinitrobenzoyl	4	135
t-Butyl	two 4-R-3,5-Dinitrobenzoyl (R=H, CH_2NMe_2, CH=CH_2	4	130–134
t-Butyl	three Benzoyl	4	110, 130, 140
t-Butyl	three Benzoyl, Allyl	4	110, 140
t-Butyl	four *p*-Nitrobenzoyl	6	112–113
t-Butyl	two Benzoyl (1,2)	6	130
t-Butyl	two Benzoyl (1,4)	6	130
t-Butyl	Camphorsulfonyl	8	48
Allyl	Benzoyl	4	109
Allyl	Tosyl	4	116
Allyl	*p*-Y-Benzoyl (Y=OMe, O-*t*-Bu, Me, Br, CF_3, CN, NO_2)	4	129
one Allyl, three H	three Benzoyl	4	140

p-Substituents Containing Hetero Atoms

CHO	Tosyl	4	140
CH=NOH	Tosyl	4	139
CH_2CHO	Tosyl	4	139
CH_2CH_2OH	Tosyl	4	139
CH_2CH_2Br	Tosyl	4	139
$CH_2CH_2N_3$	Tosyl	4	139
$CH_2CH_2NH_2$	Tosyl	4	139
CH_2CH_2CN	Tosyl	4	139
$CH_2CH_2NH_2$	*p*-Bromosulfonyl	4	199
two Acetyl, two *t*-Butyl	two 3,5-Dinitrobenzoyl	4	138
two Acroyl, two *t*-Butyl	two 3,5-Dinitrobenzoyl	4	139

Index C: Resorcinol-Derived Calixarenes